Fishing

North Carolina's

Outer Banks

Fishing
North Carolina's
Outer Banks

THE COMPLETE GUIDE TO

Catching More Fish from Surf, Pier, Sound, & Ocean

Stan Ulanski

The University of North Carolina Press CHAPEL HILL

A SOUTHERN GATEWAYS GUIDE

© 2011 The University of North Carolina Press
Fish illustrations in Chapters 9 and 10 © 2011 Duane Raver.
Illustrations can be seen in color at www.uncpress.unc.edu.

All rights reserved. Designed by Kimberley Bryant and set in Merlo, Avenir, and
Museo types by Rebecca Evans. Manufactured in the United States of America.
The paper in this book meets the guidelines for permanence and durability of the
Committee on Production Guidelines for Book Longevity of the Council on Library
Resources. The University of North Carolina Press has been a member of the Green
Press Initiative since 2003.

Library of Congress Cataloging-in-Publication Data
Ulanski, Stan L., 1946–
Fishing North Carolina's Outer Banks : the complete guide to catching
more fish from surf, pier, sound, and ocean / by Stan Ulanski.
p. cm.
"A Southern Gateways guide."
Includes index.
ISBN 978-0-8078-7207-9 (pbk : alk. paper)
1. Saltwater fishing—North Carolina—Outer Banks—Guidebooks. 2. Marine fishes—
North Carolina—Outer Banks. 3. Outer Banks (N.C.)—Guidebooks. I. Title.
SH222.N8U43 2011   639.2′2097561—dc22   2011009284

15 14 13 12 11   5 4 3 2 1

MIX
Paper from
responsible sources
FSC FSC® C013483
www.fsc.org

Contents

Illustrations, Figures, Maps, & Table

MAPS

TABLE

Preface

The Outer Banks of North Carolina are among the foremost saltwater fishing destinations, including the Florida Keys and Cape Cod, along the Atlantic coast. Drawn to their fertile nearshore waters and indigo-blue pelagic waters, anglers have descended upon this cluster of barrier islands from all over the eastern seaboard and beyond. Probably no other regions have more variety of estuarine and marine environments than these thin ribbons of sand. The Outer Banks, which are surrounded by Currituck, Roanoke, Albemarle, Pamlico, and Core Sounds to the west and the Atlantic Ocean to the east, offer a variety of fishing opportunities that are unrivaled up and down the coast.

At Cape Point, the eastern tip of Hatteras Island, prey and predator converge in the shallow, turbulent waters of Diamond Shoals. Nutrient-enriched waters ebb and flood through tidal inlets, which breach these barrier islands, and provide nourishment for juvenile and adult fish. The Gulf Stream, a warm ocean current known for its scope, range, and power, is a conduit for tropical game fish that cruise the offshore region in search of an easy meal.

This book is based on angling know-how melded with a type of information that anglers do not routinely incorporate into their fishing outings: the natural history of the waters of the Outer Banks. It is an angler's guide for targeting the major game fish that inhabit these estuarine, coastal, and blue-water environments. I systematically develop linkages among habitat, prey, and predator in this book in order to demonstrate that fishing success often depends on understanding these connections. At its core, catching fish can be boiled down to three factors: knowledge of the habits of your quarry, mastery of the necessary angling skills, and, simply, luck. But to stress the importance of the first two components, I take a holistic approach to fishing the Outer Banks, weaving angling techniques with the ecology of the fish that swim these waters and emphasizing the uniqueness of the watery environments that these species inhabit.

Fishing the Outer Banks can take many forms: casting lures to a school of blitzing bluefish, trolling for dolphinfish (mahi mahi) along sargassum weed lines, and delicately presenting a fly to a wary seatrout on a grass flat. Of the hundreds of thousands of yearly visitors to these seashores, many are avid anglers; some would say they are even fanatical in their approach to fishing. Whether it is the sleep-deprived surf fisherman who is intent in his quest for the nocturnal-feeding red drum, or the crusty boat captain who routinely barks orders to a group of novice anglers, both are passionate about fishing. The waters of the Outer Banks have seized their souls and those of many fellow anglers, so this book will offer examples of just how consuming these angling pursuits can be.

An introductory chapter on the Outer Banks allows both nonanglers and anglers alike to attain a geographical and historical perspective of the major components of this stretch of the North Carolina coast: Carova to Nags Head, Cape Hatteras National Seashore, and Cape Lookout National Seashore.

The following chapters (2 to 7) focus on particular angling opportunities: surf, pier, sound, offshore, inshore, and wrecks. Each chapter offers information on specific angling methodology; appropriate tackle; locating fish; environmental factors (tides, currents, temperatures) affecting fishing; threats to the fishery, such as the impact of beach closures on fishing; and the ecology of the area and its impact on its aquatic residents. Chapter 8 presents a brief overview of fish anatomy, emphasizing the different and unique modes of fish locomotion, prey detection, and feeding strategies among some Outer Banks game fish.

Chapters 9 and 10 detail the characteristics, biology, and behavior of thirty-five major game fish that anglers routinely target throughout both the inshore and offshore environments. These species range from small pan fish, such as spot, to blue marlin of over 500 pounds. Also included in these chapters are angling techniques and tips for each species. Each game fish is matched with an illustration to help you identify your catch. With regard to nomenclature, I have used the common name of the species that is found within the scientific literature and is widely recognized over the range of the species. For example, sea mullet, the name used by many residents of North Carolina, is listed as kingfish, the scientifically accepted common name.

The last chapter introduces anglers to the proper care of their fish, including storing and cleaning as well as how to impress your guests when you get down to some serious cooking of your catch.

The book is aimed at novice and intermediate anglers, all of whom will be able to learn something from it. Of course, this category encompasses most anglers to the Outer Banks. Even highly experienced anglers could find sections of the book useful if they wish to learn more about an area or species on which they hadn't previously spent much time or effort.

In summary, I've written this book as an aid for planning your next fishing trip to the Outer Banks as well as a reference to the aquatic ecology of the area. Take it to the beach or pier or on the boat to help you catch more fish and to enjoy the unique coastal and open ocean environments of the Outer Banks. Hopefully, if some of the information in this book allows you to hear the two words that all anglers long for—"fish on!"—then my effort in writing it will be rewarded.

Acknowledgments

Any literary work is a product of many talented people, and I would be remiss in not recognizing them.

I extend my thanks to Dr. Jeffrey Buckel of North Carolina State University and an anonymous reviewer for their very insightful and detailed comments on the manuscript. Their thoroughness in reading the manuscript significantly improved its clarity.

All game fish illustrations are by Duane Raver, noted North Carolina wildlife artist. Richard Blessing, a former student at James Madison University, used his drawing prowess to craft many of the initial figures that would ultimately be integrated into the book.

The personnel at the University of North Carolina Press, from the manuscript editorial staff to the marketing staff, are all first rate, and it has been my distinct pleasure to know many of them. A special thanks to my editor, Elaine Maisner, whose continued support is greatly appreciated.

And finally, I am indebted to the many anglers, captains, mates, boaters, and tackle shop owners whose enthusiasm for the waters of the Outer Banks and their many residents has been infectious.

Fishing
North Carolina's
Outer Banks

Outer Banks *Where Sand & Sea Meet*

South of the Virginia border are thin, sandy strands of land, known as the Outer Banks of North Carolina, which stretch more than 175 miles along the Atlantic coast. Configured like a bent arm, the Outer Banks run initially from the seaside town of Carova southeasterly to the crooked elbow of Cape Hatteras, a point of land jutting precariously into the Atlantic Ocean. From here, the Outer Banks sweep southwesterly to Cape Lookout and Beaufort Inlet, the terminus. A journey along the Outer Banks is one of island-hopping—crossing tidal inlets that separate this string of low and narrow barrier islands.

But geography alone does not tell the whole story of the Outer Banks because at their essence is their uneasy alliance with water: the brackish water of the bays and sounds that separate the Outer Banks from the mainland, the tidal currents that surge through the inlets, the ocean waves that constantly reshape and alter the beach profile, and the Gulf Stream, a powerful ocean current that brings both warmth and life to the offshore environment.

The geologic history of the Outer Banks even begins with water, or to be more exact, with ice. North America's most recent glacial period occurred in the Pleistocene epoch, which began about 1.8 million years ago and ended only 10,000 years ago when the huge continental ice sheet started to melt. Over the next 5,000 years, sea levels rose rapidly, more than 300 feet higher than during the Pleistocene. The melted water encroached over the land, inundated vast areas of the relatively flat continental shelf, and isolated a long, high ridge of sand from the mainland. Thus were born the Outer Banks, an elongated strip of land separated by more than twenty-five miles from the mainland by Pamlico Sound.

But since their birth, the islands have not been static. On the lowest-lying areas of these barrier islands, those not buttressed by protective dunes, storm waves routinely washed over the islands, carrying ocean-stirred sediments to the quiet waters of the sounds and forming large depositional flats known as overwash fans. As a long-term geological effect of this overwash, the ocean-facing shore decreased in width, but the shoreline of the sound increased in width. Though significant changes have occurred, the net result for the barrier island complex was, and continues to be, a migration toward the mainland. The islands are on the move, literally rolling over in a never-ending confrontation with the ocean.

Barrier islands, where land and sea merge, block the high-energy waves and storm surges of the ocean that would otherwise batter the coastal mainland. But barrier islands pay the price when atmosphere and sea conspire against them. Here at the ocean's edge, there are no well-defined boundaries marking where the water ends and the land begins. The islands are in a constant mode of change, and their appearance alters over time. The forces responsible for this dynamic state are among the most elemental in nature. Seasonal wave activity constantly reworks the beach sands: the gentle waves of summer gradually build up the beach, and storm waves of winter abruptly transport the sand to bars in the surf zone. The beach is in rhythm with the seasonal changes of the sea, gaining and losing sand as the year progresses. An angler successfully fishing a stretch of beach one summer day may find it unrecognizable during the winter after storms have pounded away at the fragile shoreline.

In particular, North Carolina's offshore waters are the breeding grounds for a fierce winter storm—the nor'easter—which can spin up to strength quite rapidly. Because its center of rotation is just off the coast, this intense low-pressure system pushes water from offshore to the coast. Nor'easters owe their formation to the thermal contrast between the warm Gulf Stream and the cooler coastal water. When cold, dry continental air flows out over the Gulf Stream, enormous transfers of heat and water vapor—fuel for the incipient storm—occur from the ocean to the atmosphere. These air-sea exchanges create a very unstable and volatile situation that can transform a small storm into a monstrous one virtually overnight.

Decadal-long beach studies along this stretch of the North Carolina coast have shown that 63 percent of the shoreline is eroding, 31 percent is accreting, and only 6 percent is stable. Shorelines that are thinning are vulnerable to breaching during storms. Breaching occurs when the barrier island is cut and a channel or tidal inlet forms. Even though ocean storm waves, pounding against the shoreline, may set the stage for inlet formation, seldom are barrier islands cut from their seaward side. In most cases, the breaching is the result of an elevated rise in storm water in the sound. When the level of the ocean tide falls, the elevated water rushes across the land toward the ocean, gradually cutting a channel. The present spatial distribution of inlets along the Outer Banks is related to the tidal range (the distance between

NEARSHORE FORESHORE BERM

Components of a beach (author's photo)

high and low tide). Microtidal coasts, such as the Outer Banks, have tidal ranges less than six feet and few inlets (only five between Virginia and Beaufort, N.C.).

Despite these changes, there are certain features that barrier islands have in common. A surf fisherman casting to a school of fish may have his feet firmly planted on the relatively high part of the beach—the berm— above the normal reach of waves and tides. This same angler carefully slides a hooked fish up the gradually sloping foreshore, the zone between the low and high tide levels. His only companions are mole crabs (sand fleas), scurrying about and filtering tiny pieces of organic matter called detritus. But the nearshore, extending through the surf zone, is where life abounds. Baitfish, such as silversides and menhaden, forage in the shallows and seek shelter from predatory bluefish, red drum, and flounder—a predator-prey act played out throughout the Outer Banks.

CAROVA TO NAGS HEAD

The segment of the Outer Banks from Carova to Nags Head is not an island but part of a peninsula that terminates at Oregon Inlet. This narrow strip of land, referred to by coastal geologists as a spit, results from sand deposited by waves and currents over decades.

The major movers of sand along the Outer Banks are localized currents, which like giant conveyor belts ceaselessly transport sand grains either up or down the coast. On the average, coastal currents carry a sediment load of 600,000 tons per year to the south along this northern sector of the Outer Banks. The net result is a spit approximately fifty miles long that relentlessly continues its growth southward. As this spit has grown over time, it has isolated a body of water, now known as Currituck Sound, from the Atlantic Ocean. About thirty miles long and three miles wide, Currituck Sound extends from Virginia on the north to the town of Kitty Hawk on the south.

Visitors traveling south on US 158 can cross Currituck Sound via the Wright Memorial Bridge, which links the mainland to coastal communities of the Outer Banks. At the town of Southern Shores, US 158 merges into NC 12, the main north-south highway on the Outer Banks. By heading north on NC 12, motorists can reach the towns of Duck and Corolla. Until the development boom of the 1980s, these towns were not much more than beach outposts, with little pressure from the outside world. All that has changed over the last twenty-five years; thousands of new houses have been built to accommodate the accompanying population influx. The insular nature of these communities limits fishing opportunities for the visiting angler. In particular, all three towns have no public beach access. You can't go to the beach unless you're staying in a rental cottage or bed-and-breakfast. Even those fortunate enough to obtain access find a beach that is generally wide and flat, often lacking the bottom structure that attracts prey and predator alike. Both Duck and Southern Shores do not allow off-road vehicles (ORVs) on the beach, but beach driving is permitted to a certain degree in Corolla. At the northern end of Corolla, NC 12 transitions into a sandy trail that leads to the beach. From here, anglers can only travel north to Carova; a large barrier prevents access to the southern beaches. (Check the following website for beach driving regulations from Corolla to Nags Head: www .outerbanksbeachguide.com/newsinfo/randr.htm.)

Fishing opportunities increase dramatically if one heads south on NC 12. Though this part of the Outer Banks is the most heavily developed, with mile after mile of shopping centers, motels, waterslides, restaurants, and beach cottages, the towns of Kitty Hawk, Kill Devil Hills, and Nags Head cater to the angler.

Each town has numerous public access sites within its boundaries. These

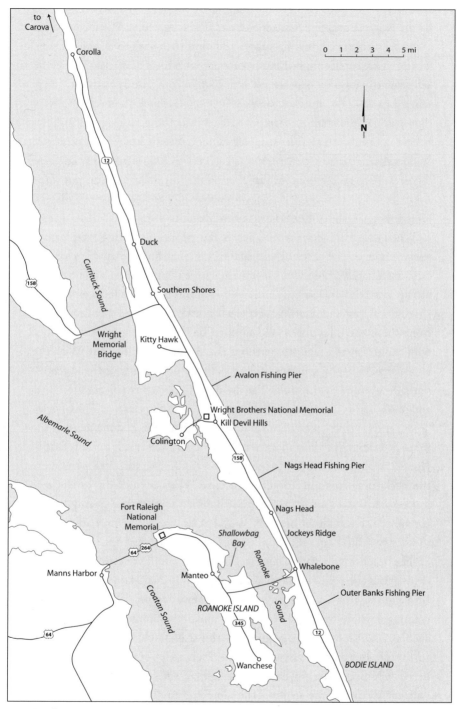

to
Carova

Corolla

12

Duck

158

Currituck Sound

Southern Shores

Wright
Memorial
Bridge

Kitty Hawk

Avalon Fishing Pier

Albemarle Sound

Wright Brothers National Memorial
Kill Devil Hills

Colington

158

Nags Head Fishing Pier

Fort Raleigh
National
Memorial

Nags Head

Jockeys Ridge

Shallowbag
Bay

64 264

Manns Harbor

Manteo

Roanoke Sound

Whalebone

Outer Banks Fishing Pier

Croatan Sound

ROANOKE ISLAND

345

64

Wanchese

12

BODIE ISLAND

0 1 2 3 4 5 mi

N

Corolla to Nags Head

sites fall mainly into two groups: neighborhood and local, as categorized by the North Carolina Division of Coastal Management. The former offer a limited number of parking spaces and sand pathways to the beach. The local sites are often simply dune crossovers, with little or no parking. These are primarily used by pedestrians who own or rent nearby cottages. In essence, beach access requires the existence of clearly marked rights-of-way that may pass through private property. While these access points offer a boon to persons not fortunate or rich enough to own beachfront property, anglers should be aware that not all beach access points lead to prime fishing sites. (The access sites are available on the following website: www.the coastalexplorer.com/NC/KittyHawk/community.asp. Substitute KillDevil Hills or NagsHead for KittyHawk for the other towns.)

When they visit their favorite beach, few people, including even experienced anglers, worry about their legal rights regarding the use of this property. Fortunately, those who assume they have the right to swim, boat, and fish are correct. The legal grounds on which this right is based extend back to medieval England. During this period, the king declared that those lands over which the tides ebbed and flooded be held in trust for his subjects, who could fish and navigate without the threat of punishment. With the establishment of the American colonies, this "public trust" doctrine was essentially adopted by the states. The dividing line between private shorelines and public trust lands came to be the mean of all high tides over a lunar cycle of 18.6 years. While I won't bore you with the problems of documenting the exact position of the mean high-water mark, the important point is that in North Carolina, recognition of this high-water line is, and always has been, the legal ruling. But don't forget you license. Whether you are a North Carolina resident or a visitor to the Outer Banks, a recreational fishing license is required to fish either ocean or sound. A license can be procured online (www.ncwildlife.org) or from some Outer Banks tackle shops.

The beaches from Kitty Hawk to Nags Head offer a variety of surf species throughout the year. Their availability depends primarily on water temperature and the abundance of prey. As sea surface temperatures rise with increasing daylight during the late spring and early summer, tropical species, such as Spanish mackerel, migrate northward and chase silversides in the surf break. But these same beaches also receive a lot of pressure from other users, including surfers, swimmers, sunbathers, shell collectors, and joggers, particularly during the height of the summer tourist season. Many seasoned anglers avoid the crowds by concentrating their efforts during the morning and evening hours. Even the fish tend to be more cooperative at these times, feeding more vigorously than during the heat of the day. With cooling water temperatures later in the year, northern species, such as striped bass, delight surf casters. Having fattened up during the summer, striped bass migrate southward to their winter haunts.

In addition to beach access, the infrastructure of the towns supports fishing. There are three ocean piers (see Appendix 1) that jut hundreds of feet into the Atlantic. Whether your interest lies in catching small bottom-feeders, such as spot and croaker, or zeroing in on speedy cruisers, like king mackerel, opportunities abound on the piers. Piers are also a great place to start young children fishing because they can use very simple tackle to catch some interesting and, at times, abundant species, like pigfish and pinfish, which are generally not sought-after by other anglers.

Tackle shops (see Appendix 2) in this area not only stock the necessary hardware (rods, reels, and lures) and bait (menhaden, mullet, shrimp, and squid) for the angler but also are sources of advice. In addition to the knowledgeable personnel offering information about appropriate tackle and bait needed to catch a particular species, many of these establishments have websites with daily postings about what is biting, where it is biting, and how to get the bite.

If your angling tastes tend toward the quiet backwaters behind the beaches rather than the roaring surf, both Roanoke and Croatan Sounds offer a wide variety of light-tackle fishing opportunities. The sounds are separated by Roanoke Island, which in 1585 was first settled by a band of hardy English colonists. The settlement was short lived, and historians now refer to it as the "Lost Colony."

While much of the sounds' shorelines is privately owned, and thus not easily accessible to the wading angler, the public has some rights. In one litigated case (*State v. Twiford*), a property owner blocked off Jean Guite Creek in Currituck Sound, claiming she had sole ownership of the creek bottom. In response, the court studied historic use of the creek as a shelter from storms and thoroughfare for the public and proclaimed that even if the land were in private hands, the public would have an easement to pass by boat over the water. From an angler's perspective, this ruling means that even private owners of small creeks, marshlands, and tidal flats cannot prevent or inhibit public access by boat.

The Outer Banks have a variety of marinas (see Appendix 3) offering easy access to sound, inshore, and offshore fishing. With regard to the latter, hundreds of anglers each year book a charter to tackle trophy game fish, some weighing more than 500 pounds, which swim the Gulf Stream waters. Almost daily, dozens of diesel-powered sportfishers reaching lengths of more than fifty feet strike out for the blue offshore water. In contrast, other anglers opt to hire a guide specializing in shallow-water fishing, targeting species that roam the sounds, inlets, and nearshore environments. Some of the marinas even have boat-launching ramps for anglers willing to seek out the fish on their own. Pirates Cove Marina and Broad Creek Fishing Center, both located on Roanoke Island, begin a line of marinas strung out along the Outer Banks.

CAPE HATTERAS NATIONAL SEASHORE

On January 13, 1953, a seventy-two-mile strand of sand in eastern North Carolina stretching from South Nags Head to Ocracoke Inlet was designated the Cape Hatteras National Seashore—the first national seashore in the United States. This designation put to rest a sixteen-year battle over the use of this land. Though Congress first authorized establishment of a national seashore in 1937, funds were not appropriated for its acquisition because of opposition from land developers and oil companies that hoped to reap an economic windfall from this coastal environment. Finally in 1952, Paul Mellon and his sister donated $800,000 to purchase the land, and when the State of North Carolina matched this gift, the land was protected from private speculators and a national seashore became a reality. Except for preexisting villages, the 30,000 acres of the seashore are in the public domain and are managed by the National Park Service (NPS).

The seashore encompasses some of the most historic and environmentally fragile real estate in the United States. As early as the sixteenth century, Sir Richard Grenville, who was commissioned by Sir Walter Raleigh to establish an English colony in the New World, explored its marshes, sounds, and inlets. More than a hundred years after the first European had set foot in coastal North Carolina, the pirate Blackbeard, who served as a privateer in Queen Anne's War (1702–13), seized merchant and slave ships from his base on Ocracoke Island. The Cape Hatteras Lighthouse, which was built in 1870 as a welcoming beacon to mariners navigating the treacherous Diamond Shoals, was moved inland in 1999 as a concession to the eroding shoreline.

Though numerous markings, brochures, and postings commonly refer to this stretch of windswept coast as the Cape Hatteras National Seashore, legally (U.S. Code: Title 16,459) it is officially titled the Cape Hatteras National Seashore Recreational Area. One of the main recreational activities enjoyed by thousands of visitors to this land and the surrounding waters is fishing. Over the decades, numerous world records, as documented by the International Game Fish Association, have come from the seashore's waters, including a thirty-one-pound bluefish and a ninety-four-pound red drum. Let's take a tour of the area to get our bearings.

The northern entrance to the seashore starts at Whalebone Junction, the intersection of NC 12 and the highway west to Roanoke Island. Heading south on highway 12, the vista changes abruptly from rampant development to a landscape of salt-resistant trees and shrubs struggling to keep a hold in the shifting sands. Historians refer to this stretch of land as Bodie (pronounced body) Island; however, it is not a true island because Roanoke Inlet, which originally separated this section from land to the south, has long since closed. (Over the last 500 years, more than two dozen inlets have at times dotted the Outer Banks.) Still standing guard is the Bodie Island Lighthouse, which was built in 1872. This 170-foot structure is the third lighthouse to bear the name Bodie Island. The first lighthouse, built in 1848, succumbed to structural damage in only ten years, and the 1859 lighthouse was destroyed by Confederate forces who were determined that it not fall into Union hands.

The first of numerous pedestrian access points to the Atlantic side of the seashore is Coquina Beach—a swimming, fishing, and general recreational area. The beach derives its name from the tiny, butterfly-shaped coquina clams that burrow into the sand at the water's edge. These edible clams are not only a common ingredient in chowder, but local artisans employ the shells in their particular crafts, such as brick making.

At the south end of Bodie Island and approximately five miles from the entrance is Oregon Inlet, which is the only outlet to the ocean in the 140 miles between Cape Henry in Virginia and Hatteras Inlet. Oregon Inlet is the main passage into or out of Pamlico Sound for commercial and recreational fishing vessels based along the northern part of the seashore. Since its opening in 1846 during a hurricane, Oregon Inlet has migrated approximately two miles to the south of its original location as the result of the southward-flowing current. Even today, the inlet is in a dynamic state of change and poses a navigational test to even the most experienced captains. Since 1960 at least thirty lives and an equal number of boats have been lost

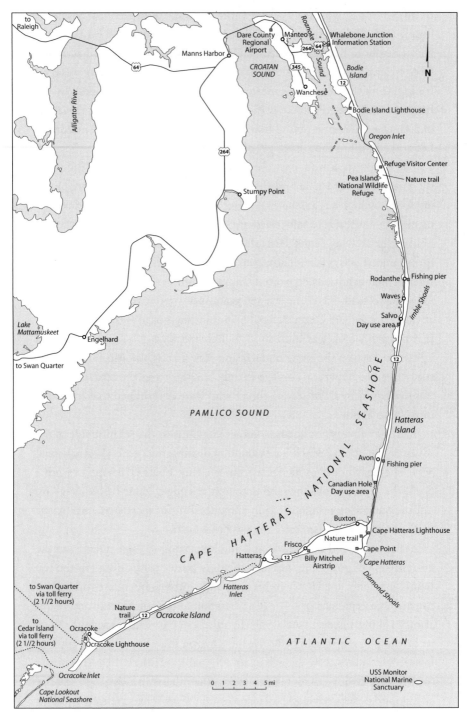

Cape Hatteras National Seashore

due to a combination of the inlet's swift tidal current and shifting shoals. In order to maintain a relatively passable channel, dredging of the inlet by the Army Corps of Engineers occurs throughout the year at a cost of $10 million. While to the casual observer this appears to be an exorbitant amount for moving sand around, according to a relatively recent in-depth study, the effect of losing Oregon Inlet as a navigable waterway for commercial and private boaters could be economically devastating. The overall losses could reach more than $682 million and directly or indirectly affect over 9,600 jobs.

Ramp 4, just north of Oregon Inlet, is one of sixteen ramps in the seashore that facilitate vehicular access to the Atlantic side. (Ramp numbers refer to the approximate miles on NC 12 south of Nags Head.) But be aware that only four-wheel-drive vehicles should attempt to traverse the soft and deep beach sands. For efficient and safe navigation, anglers should reduce tire air pressure (the NPS recommends twenty psi) on their vehicles and carry the following equipment: good spare tire, working jack, wide board to support the jack, and tow rope. (Employing a commercial towing service to extract a stuck vehicle can be quite expensive.) The NPS reserves the right to close stretches of the beach to ORVs due to storm erosion or concern about endangered wildlife. In addition, Ramp 2, which is near Coquina Beach, is closed seasonally between May 15 and September 15. (From April to October the NPS issues weekly beach access reports, which can be accessed on the following website: www.nps.gov/caha.) To an avid surf angler, a well-equipped ORV is an efficient means of transporting gear to a favorite fishing spot. The NPS strictly enforces the beach speed limits and severely frowns on driving over dunes or through restricted areas.

The Herbert C. Bonner Bridge, which spans Oregon Inlet, connects Bodie Island to the Pea Island National Wildlife Refuge. On the south side of the bridge is an ocean-side catwalk that is accessible from the adjacent parking lot. Extending hundreds of feet into Oregon Inlet, the catwalk allows shorebound anglers to fish the deeper waters of the inlet. In particular, fishing for structure-oriented species, such as sheepshead and striped bass, can be quite fruitful depending on the time of the year.

Pea Island, which derives its name from the wild pea vine that grows there, encompasses 6,000 acres of land and almost 26,000 acres of the bordering Pamlico Sound. It is managed by the U.S. Fish and Wildlife Service as a resting and wintering site for ducks, snow geese, and other migratory birds. Like Bodie Island, Pea Island is not a true island, but it is a section (thirteen miles) of Hatteras Island, the first of the two barrier islands (Ocracoke is the other) within the seashore. Hatteras Island curves out into the Atlantic and then arches back in a protective embrace of North Carolina's mainland and Pamlico Sound. In contrast, Ocracoke Island stretches southwesterly for approximately fourteen miles.

While ORVs are forbidden in the Pea Island preserve, anglers can try their luck throughout the refuge by either pulling over to the side of NC 12 and walking over the dunes (tread lightly) to the Atlantic side or utilizing a number of designated parking spots to reach the beach or Pamlico Sound. The refuge is open to recreational activities only during daylight hours; however, anglers can obtain permits from the park service's main office in Manteo to surf fish at night (September 15 to May 31).

The extensive network of ocean-side dunes, which are an integral part of the barrier islands' ecosystem, did not develop naturally. During the Great Depression, the Civilian Conservation Corps set up camps along the barrier islands with the purpose of dune construction and shoreline stabilization. Since these initial efforts in the 1930s, the NPS has waged an almost constant battle to maintain the structural integrity of the dunes. Because dunes act as flexible barriers to storm surges and help protect these low-lying islands, the NPS plants and replants a variety of perennial grasses and shrubs that can tolerate the harsh growing conditions on these salt-encrusted coasts. One of these rugged species is sea oats that are easily recognized by their tall size (three to four feet) and slender shape. Because of their extensive root system, sea oats are effective trappers of sand and can colonize a barren dune in a few growing seasons.

Immediately south of the refuge is the village of Rodanthe. By 1850, thirty-seven hardy families had settled here and scratched out a living by grazing cattle, raising vegetables, and fishing the fertile waters. Similar to the other towns in the Cape Hatteras National Seashore, Rodanthe is an enclave of privately held lands nestled within the seashore. Rodanthe merges with the towns of Waves and Salvo to form a densely developed strip of land that stretches for approximately five miles south on Hatteras Island. Though tourism is now the main industry, long ago having replaced commercial fishing, recreational anglers are not totally shut out. At approximately milemarker 20 is the Cape Hatteras Island Fishing Pier, one of two piers on Hatteras Island.

The proliferation of vacation homes, campgrounds, and shops in this area limits access to both surf and sound. Also, ORVs are not permitted on the beach during the height of the summer tourist season. Any serious surf or bay angler would be well advised to continue heading south back into the seashore.

From Ramp 23 to Ramp 34, which is just north of the town of Avon, is one of the longest stretches within the park of readily accessible ocean fishing. Each of the four ramps (23, 27, 30, and 34) has parking for those lacking ORVs and willing to lug their gear to the water's edge. While theoretically one could drive the total eleven miles, access is often limited due to frequent beach closures. In addition, during periods of abnormally high tides, stretches of the beach may be impassable. Though not as clearly marked as the ocean-side ramps, there are a few sandy trails leading to Pamlico Sound.

A few miles south of Rodanthe is the town of Avon, which was originally known by the Indian name of Kinnakeet. Over time, vacationers renting upscale beach houses have replaced the Native Americans. Though Avon is essentially bounded on the north by Ramp 34 and on the south by Ramp 38, the beach in between is closed completely to vehicular traffic during the summer, and the NPS routinely limits access even during the off-season.

South of Avon, the island narrows appreciably, and the sea routinely encroaches on the highway. During the Ash Wednesday Storm of 1962, the dunes were breached, the road was cut, and a new inlet formed. Over time, the inlet was filled and the road was rebuilt. Inlet formation due to storms has been and will continue to be a recurring theme on these barrier islands. While the opening and closing of inlets can cause major disruptions for the seashore's residents and visitors, inlet dynamics are part of the natural system on barrier islands, facilitating the vital mixing of salt water with fresh water in the sound.

Approximately six miles south of Avon is the town of Buxton, another enclave of private lands. Buxton is located on one of the widest and most stable sections of the Outer Banks. In the sixteenth century, this part of Hatteras Island was called "Croatoan" by the colonists of Roanoke Island. The Croatoan Indians, the cape's earliest inhabitants, were later known as the Hatteras tribe, from which the cape derives its name. The easternmost tip of Hatteras Island, which is known as Cape Point, can be reached by turning off of NC 12 onto a park service road that winds past the Cape Hatteras Lighthouse and Buxton Woods, one the largest maritime forests along the Atlantic coast. After a few miles, one comes upon an NPS campground and the first drive-over (Ramp 44) to the beach. Just a bit to the north is another ramp (Ramp 43). Ocean overwash during storms often floods these ramps, resulting in closures by the NPS. Though anglers can access Cape Point by walking, it is a considerable distance from the ramps, and this hike is for those anglers willing to travel lightly. The very tip of the point, a large, barren sand flat, is constantly being shaped and changed by the almost encircling sea. Waves converge and shoot water skyward in the shallows known as Diamond Shoals, the submerged tail of the cape, which stretches seaward for several miles. If you park at the tip of Cape Point, be mindful of the tides. A few unwary anglers have experienced the sinking feeling of seeing their vehicles swamped by the encroaching water.

Southwest of Cape Point is the community of Frisco and Ramp 49, which allows access to an area known by the locals as "south beach," a concave shoreline on the edge of Buxton Woods. Anglers can also access south beach by using Ramp 44 and following a sandy, inland trail that veers off to the right of the ramp's trial.

Beyond Frisco, the island again narrows, and Sandy Bay, an extension of Pamlico Sound, cuts deeply into the island. In 2003, hurricane Isabel breached this section of Hatteras Island. Though the new inlet was ultimately filled and this segment of NC 12 was rebuilt, coastal geologists worry about the long-term stability of this part of the island.

Beyond Sandy Bay is the village of Hatteras, with deep roots in commercial fishing. But for the recreational angler, Hatteras's main attraction is its four marinas that specialize in offshore fishing. Each afternoon, crowds gather at the docks to watch the boats unload their catch—a social gathering marked by happy anglers mingling with family and friends.

South of Hatteras is Ramp 55, which allows access to the southern tip of Hatteras Island that terminates at Hatteras Inlet. Anglers have the option of driving along the beach or utilizing a sandy, inland trail, which winds past an old lookout tower, to reach the inlet. This back trail is, however, subject to periodic weather and resource closures. Essentially opposite Ramp 55 is a free, state-run ferry service that links the islands of Hatteras and Ocracoke. Though the ferries make the crossing on a frequent and regular basis, long lines to board them are common during the summer. The ferries accommodate thirty vehicles, including cars, campers, and recreational vehicles, and the trip across Hatteras Inlet takes about forty minutes.

Except for the town of Ocracoke, which is located at the south end of Ocracoke Island, the island is known for its wild, uncrowded beaches that offer miles of surf fishing opportunities. In 2007, Stephen Leatherman, known as "Dr. Beach," ranked Ocracoke Island as his number one beach destination in the United States.

Ocracoke Island has changed dramatically in appearance since colonial times, a befitting example of the dynamic nature of this part of the North Carolina coast. An 1585 map depicts an island about eight miles long with the original Hatteras Inlet cutting right through the middle of the present island. Over time, this old inlet closed, and by the 1750s Ocracoke was attached to Hatteras Island until the current Hatteras Inlet opened in 1846.

Today Ocracoke is a typical barrier island—long, narrow, covered with low shrubs, and fronted by a dune line. Five ramps (59, 67, 68, 70, and 72) allow access to almost this entire fourteen-mile strip of sand, but anglers should check about seasonal closures. Ramp 59 is near the south side of Hatteras Inlet, and Ramp 72 allows anglers to reach Ocracoke Inlet, which is the only inlet on the Outer Banks to have remained open through recorded history.

Ocracoke Island has no piers, but it does have a marina that is located in the village of Ocracoke. Ferry service from Cedar Island and Swan Quarter on the North Carolina mainland carries both passengers and vehicles to Ocracoke Island.

CAPE LOOKOUT NATIONAL SEASHORE

After crossing Ocracoke Inlet, one enters a realm of wide, bare beaches, low dunes covered by windblown grasses, barrier flats bordered by dense vegetation, and large expanses of sound-side salt marshes. It is the wildest and most isolated portion of the Outer Banks, standing in sharp contrast to the intensively developed and densely populated northern Outer Banks. Most of the Outer Banks must have looked like this when English colonists first set foot on this land in the sixteenth century. This is the Cape Lookout National Seashore, a string of barrier islands (Portsmouth, Core Banks, and Shackleford Banks)—fifty-five miles in length—stretching from Ocracoke Inlet on the northeast to Beaufort Inlet on the southwest, with two inlets in between. The islands are separated from the mainland by the shallow Core and Back Sounds, which have depths of less than one foot at low tide over approximately 40 to 60 percent of their area.

Though designated a national seashore by Congress in 1966, the Cape Lookout islands have been spared any government attempt to modify them.

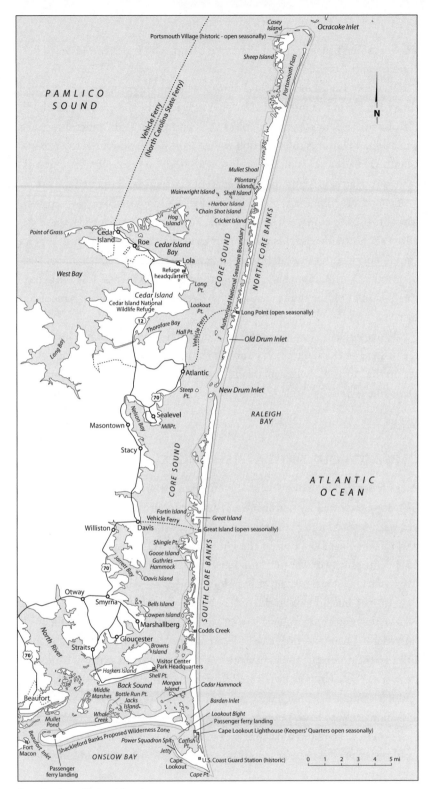

Cape Lookout National Seashore

The artificial dune system, prevalent on the Cape Hatteras seashore, is absent, as are any roads or bridges. In place of the asphalt thoroughfares are the ephemeral sandy tracks of ORVs, and boats of all styles and sizes are the only means of accessing these islands. And yet historians have documented periods of occupation by Native Americans and early settlers. The name Core Banks derives from the Coree Indians who once lived on the mainland but hunted on the Banks.

A 1590 map of the North Carolina coast labels present-day Cape Lookout as "promontorium tremendum"—horrible headland—in recognition of the cape's treacherous shoals. These shoals, along with those off Cape Hatteras, have claimed hundreds of unwary ships and mariners over the centuries, giving the coast of North Carolina the dubious distinction of being known as "the graveyard of the Atlantic." Despite rising sea levels, coastal geologists have determined that the cape and shoals have accumulated vast quantities of sediment due to converging currents, in contrast to the gradually retreating barrier island limbs. For example, the Core Banks, to the north of Cape Lookout, have eroded on the oceanfront at the rate of 165 feet per century. Experiencing the full force of winds and storm surges, the low dunes are regularly overwashed. In response, the sound-side marshes have grown, and the islands have retained their shape and form, retreating to the mainland.

No coastal features on the Outer Banks are ever static; beaches erode, shoals shift, and salt marshes grow. The operative word is "dynamic," and so it is with the inlets of the Cape Lookout National Seashore. Over the centuries, the seashore's inlets have regularly opened and closed in response to the volatile nature of North Carolina's offshore waters. In particular, Old Drum Inlet on the Core Banks first opened in the eighteenth century and again in 1933, before closing in 1971. At present there are three open inlets, Drum (created by U.S. Army Corps of Engineers just south of Old Drum Inlet), Barden, and Beaufort, along this section of the North Carolina coast. Two inactive inlets, Whalebone and Swash, on the northern Core Banks, periodically open and close.

Accessing the seashore can be an adventure. While private boats make the short crossing (three miles) across Core Sound to the sound-side marshes, there are flotillas of ferries that routinely make the journey from the mainland to the seashore. Some of the smaller, flat-bottomed ferries carry only passengers, while the larger vessels carry both passengers and vehicles. In general, only ferries to the Core Banks, the major portion of the Cape Lookout National Seashore, carry vehicles. The NPS maintains a list of authorized ferry services (www.nps.gov/calo/planyourvisit/ferry.htm).

At the northern end of Portsmouth Island is the village of Portsmouth, reachable by public ferry from Ocracoke. Portsmouth village is now abandoned but was a thriving seaport community during the nineteenth century. It sits on the widest part of the national seashore, more than two miles wide, but erosion is taking its toll as Ocracoke Inlet migrates to the south. A stroll across this section of the island from the sound traverses wide salt marshes, a maritime forest, and a huge sand flat, which probably owes its origin to the large sand deposits built up around Ocracoke Inlet.

The ferry from Atlantic, N.C., allows access to the North Core Banks, and the ferry from Davis, N.C., disembarks passengers and vehicles on the South Core Banks. These two barrier islands are similar in appearance—low, narrow, and sparsely vegetated—with very little change in habitats from one end to the other. To paraphrase Dirk Frankenberg from his book *The Nature of North Carolina's Southern Coast: Barrier Islands, Coastal Waters, and Wetlands* (1997), once you've seen fifty feet of the islands, you've seen it all. The homogeneity of the area should not discourage anglers from exploring this vast natural recreational preserve and discovering their own favorite spot.

The South Core Banks stretch for approximately twenty-five miles from New Drum Inlet on the north to Cape Lookout on the south. The Cape Lookout Lighthouse, with its distinctive black-and-white diamond pattern, provides a picturesque backdrop for surf casters. First lit on November 1, 1859, the lighthouse's white light is visible nineteen miles out to sea. The Cape Lookout Lighthouse is the southernmost of a string of lighthouses (Ocracoke, Hatteras, Bodie, and Currituck) that were built along the Outer Banks approximately forty miles apart so that as soon as ships lost sight of one light, the next would come into view.

The Core Banks are the realms of the hard-core anglers. With miles of open shoreline and relatively little fishing pressure, compared with the northern Outer Banks, the angling opportunities are endless on the Core Banks: red drum in the surf, flounder in the inlets, and speckled trout in the sound. At Long Point on the North Core Banks and Grand Island on the South Core Banks the NPS seasonally provides rudimentary services, such as restrooms, showers, and drinking water. In addition, rental cabins are available on both barrier islands (www.nps.gov/calo/planyourvisit/lodging.htm).

South of the Core Banks is the last barrier island within the seashore—Shackleford Banks, named for the man who first purchased land here in 1714.

National Park Service cabin on the South Core Banks (author's photo)

The island stretches nine miles west from Barden Inlet to Beaufort Inlet and has a geological and ecological setting distinctively different from that of the Core Banks. Because of its east-west orientation and the predominant wind direction, overwash is at a minimum; instead, sand is transported to the west, resulting in an island that is growing into Beaufort Inlet. In the 1860s, Mullet Pond on the western end of Shackleford was a saltwater bay, but by the 1960s the accreting island sealed it off, creating a freshwater marsh.

Although Shackleford Banks barely exceeds 2,000 acres, the island exhibits considerable plant diversity, including sparsely vegetated dunes, marshland, and woodland. According to a U.S. Coast and Geodetic Survey map in 1853, the island was almost completely covered by a maritime forest. But by 1899, most of the forest was gone, the result of a severe hurricane that struck and submerged the island. Today, less than 5 percent of the island is covered by forest; the western end of Shackleford contains remnants of this oak, pine, and cedar forest.

As you will soon see in the following chapters, the waters of the Outer Banks are a magnet for many game fish, some drawn to the turbulent surf, others to the fertile sound, and still others to the cold depths. But depending on the time of year, many of these species can be found in more than one environment, providing multiple opportunities for any angler.

THREATS TO THE OUTER BANKS

Though a burgeoning permanent and seasonal population has stressed and negatively affected the natural resources of these islands, Mother Nature may have the final say as to the future of this fragile ecosystem. A 2004 study conducted by geologists at Louisiana State University came to a somber conclusion: the Cape Hatteras area ranks in the top three sites (also included were southern Florida and the Gulf coast between the Florida panhandle and east Texas) for a hurricane to make landfall, and the frequency and intensity of these storms appears to be increasing. The coast of North Carolina can expect to be hit by a hurricane once every four years.

The Outer Banks are nothing more than sandbars poking through the sea surface and are less than a half-mile wide in many spots. A ten- to fifteen-foot storm surge, associated with a strong hurricane (category 4 and above), has the potential to flood these islands, causing them to disappear overnight. The turbulent history of the Outer Banks is destined to repeat itself in the near future. The beach as you now know it may no longer exist.

Surf Fishing

It has been said that no form of angling can match the thrill and excitement of surf fishing, as waves crash against the shore, marauding predators chase bait into the shallows, and long, arching casts reach far-off fish. And it has also been said that the Outer Banks rank as one of the premier surf fishing sites along the Atlantic coast. That the number of surf fishermen has grown dramatically over the years seems to support these claims. When schools of bluefish, red drum, and other species churn the nearshore waters, anglers line the beaches, sometimes shoulder to shoulder, to cast their baits and lures.

Traditionally, surf fishermen, most of whom were the permanent barrier island residents, either walked or rode pony carts from the villages to the beaches. (On Ocracoke Island, the locals routinely used ponies, which roamed free on the island, for both recreation and work.) Before World War II, the islands were isolated and accessible only by boat; thus visitation was light. But two factors would spur the growth of surf fishing: the introduction of motorized vehicles and improved infrastructure. With regard to the former, the first vehicles had to be ferried to the islands and, as recently as the 1950s, required no license plates. But surf anglers were quick to employ these vehicles as "beach buggies" that significantly increased their mobility and provided them with access to areas heretofore difficult to reach. These first vehicles, which were equipped with oversized and underinflated tires to navigate the soft sand, ultimately evolved into sophisticated and high-powered off-road vehicles (ORVs). On the second point, the entire length of Cape Hatteras National Seashore was paved in 1954, and by 1963 the Herbert C. Bonner Bridge had been built across Oregon Inlet and eliminated the need for the time-consuming ferry service. Thus the trip of six or more hours to the village of Buxton from South Nags Head was reduced to less than an hour. With these changes have come inevitable consequences. In 1941 there were fewer than a hundred beach buggies along the Outer Banks, whereas today there are tens of thousands of such vehicles that routinely traverse the beaches. To some surf anglers, solitude has become a rare commodity. Only the islands of the Cape Lookout National Seashore have remained virtually unchanged over time and thus have not received the angling pressure that the northern islands have endured. Isolation can be a true blessing.

TARGETING FISH: TACKLE AND TECHNIQUES

There are a variety of fish that swim within the cast of a surf angler. The spectrum of sizes, speeds, and feeding habits of these fish necessitates a variety of rods and other tackle. While I have seen numerous ORVs carrying a dozen or more rods, your arsenal need not be as extensive. You can cover most surf fishing situations with three rods that are designated by their respective lengths and ability to cast a certain weight or lure.

A good start for most anglers who are targeting red drum, big bluefish, and striped bass is a rod in the ten- to twelve-foot range that is capable of casting a large chunk of bait and four- to six-ounce sinkers a distance of about 200 to 250 feet and, at times, even farther. A long rod has a distinct mechanical advantage over a short rod in achieving long casts because the tip of the longer rod sweeps out a greater distance during the casting motion (see casting section below) than does the shorter rod. This action results in a longer period of accelerating and loading the rod through the casting stroke—both necessary for those desired long casts. Loading the rod refers to the potential energy stored in the bend of the rod that is released when the rod uncoils during the cast. The rod is bent by the weight (sinker or lure) when the rod is started forward in the casting motion.

In addition to the importance of the length of the rod in achieving distance, the stiffness of the rod now enters the picture. A supple rod will subtract from the distance because the rod will bend excessively as the force builds up at the rod tip to propel the weight forward. The rod simply cannot be loaded properly. In contrast, a relatively rigid rod will not bend as much during the application of the force, but it increases the efficiency of the casting stroke by bending just enough to load the rod to its maximum specification. In summary, the rod acts as a lever, albeit a complicated one, transferring energy supplied by the caster and storing some of the energy. Many anglers who target big red drum often use "Hatteras heavers," long rods in the twelve- to thirteen-foot range, capable of tossing an eight- to ten-ounce sinker and a large chunk of bait.

A shorter rod, generally nine feet, can serve a dual purpose: cast small bottom rigs (two to three ounces) that target smaller species, like flounder, kingfish, and pompano, and propel lures (one to two ounces) to Spanish mackerel and bluefish that are intent on chasing down fleeing baitfish. Being relatively light in weight, this rod allows the angler to cast repeatedly without quickly tiring.

The shortest rod, in the seven-foot range, will comfortably cast small lures (less than a half-ounce) and jigs (about a quarter-ounce) to species, like seatrout, that swim and forage at times close to the shore. This rod simply does not have the length or stoutness to cast long distances. While the range of surf species that can be targeted with this rod is limited, it can also be used for pier or sound fishing.

Choosing a reel to match your rod often comes down to a matter of preference and your proficiency in casting. Many seasoned surf anglers will match their longest rod with a conventional reel that they load with fifteen- to twenty-pound test line. With this type of reel, the spool rotates when the line is pulled off the spool due to the force of the cast. The advantages of conventional reels are threefold: heavier line can be packed on the spool, there is increased torque for playing large fish, and the angler can make longer casts. With regard to the latter, obviously the skill of the caster comes into play, but since the spool rotates during the cast, the line "flows" off the spool effortlessly. The downside of these reels, particularly for the novice angler, is the learning curve, albeit not too steep, required to master the nuances of the rapidly rotating spool. Spool rotations reach thousands of revolutions per minute, and if the angler cannot control the spool's speed, the result is the dreaded "bird's nest"—a tangled mass of line embedded into the spool. Fortunately, reel manufacturers have come up with ways to control the rotational speed and have made conventional reels more user-friendly.

Suffice it to say that the angler's other option, a spinning reel, is more than capable of both casting a heavy weight and fighting a big fish. With a spinning reel, the spool does not rotate during the cast. Because the line is pulled from the spool in a series of loops by the sinker or lure, air resistance and friction, which are caused by the loops traveling through the rod's guides, reduce casting distance. But on the upside, spinning reels are generally easier to use than conventional reels because massive tangles are essentially nonexistent.

Line twist is inevitable with a spinning reel, particularly after repeated casts. The twisting not only compromises the integrity of the line but significantly affects casting distance. The problem can be minimized by not reeling when a fish is taking line and manually closing the bail after a cast.

A nine-foot rod and a spinning reel loaded with ten- to twelve-pound test line is the preferred outfit when the angler must cast repeatedly in order to catch the attention of a fast-swimming predator. In the heat of the moment and with adrenalin pumping, even experienced anglers often make casting mistakes. The spinning reel is just more forgiving than the conventional reel in this situation. However, if you feel up to it, by all means experiment with a small conventional reel.

There is no debate about matching the light, seven-foot rod with a small spinning reel that is capable of holding 100 yards of six-pound test line. Remember that you will be casting light lures or jigs that are simply not heavy enough to initiate spool rotation on a conventional reel.

Enter any tackle shop, peruse any fishing catalog, and you will quickly note the myriad sinkers, hooks, lines, lures, plugs, and jigs that are available in all sizes, shapes, weights, and colors. How to choose? My advice is to keep it simple, limit your selection, and use those rigs that over time have proven successful in these waters. Seek advice—we're all novices at one point.

When fishing for red drum, though not limited to this species, many anglers will use a fish-finder rig. Its main feature is that the heft of the sinker does not spook the fish, since this weight rides freely up the line when the fish picks up the bait. Relatively recently, circle hooks (6/o to 9/o), rather than J-shaped hooks, have become an integral part of this rig. Circle hooks have the advantage of becoming lodged in the corner of the fish's mouth after hookup, preventing the fish from deeply swallowing the offering. In contrast, deep hooking may result in the death of a released fish, which negates the purpose of size and bag limits. Baits include chunks of mullet or menhaden.

Fishing for smaller bottom fish, such as pompano and kingfish, necessitates scaling down your rig; smaller sinkers and hooks are in order. While many anglers simply use a smaller hook on a fish-finder rig, a two-hook or even three-hook bottom rig is generally more effective and holds more bait. Multiple hookups are thus possible, particularly, with kingfish that generally swim in relatively large schools. Anglers may employ a variety of natural baits, including pieces of squid, shrimp, bloodworms, and sand fleas.

No area probably generates more debate within the surf fishing community than which lure to toss to bluefish or to Spanish mackerel that are actively feeding on small baitfish. If you choose a lure that mimics what the game fish is feeding on and is aerodynamically shaped for casting, you will generally have your share of hookups. Metal lures, such as Kastmasters,

Hopkins, and Stingsilvers, in the one- to three-ounce range have all proven successful in catching myriad species over the years. Many of these lures come in a variety and combination of colors, and all have their advocates among Outer Banks anglers. While many fish can indeed perceive color, there is no hard evidence that supports one colored lure over another in consistently catching fish. You can't go wrong with silvered lures that reflect sunlight well, provide that attractive "flash," and match in coloration some of the major forage species, like Atlantic silversides, of bluefish and Spanish mackerel.

Your tackle box should include both bucktails and jigs. These leadhead lures, which are adorned with a synthetic dressing on their hook, are deadly on some bottom-feeders. A one-ounce bucktail, slowly retrieved along the bottom, often proves irresistible to a flounder that is waiting patiently to ambush its prey. Many surf anglers who target spotted seatrout use quarter- to half-ounce jigs. When bounced along the bottom, the undulating motion of the curly, rubber tail on the hook often entices a strike from trout that are feeding relatively close to the shore.

Casting is a fundamental skill that every surf angler should develop. You simply can't catch fish if you can't place your bait or lure where the fish can see and/or smell it. All anglers should concentrate on technique—developing an efficient casting stroke that allows consistent placement of an offering in the target zone. Accuracy over distance should be an angler's top priority. Once you achieve confidence in your casting style, you can add long casts to your repertoire.

Since there are many excellent books that provide detailed instructions and photographs for a variety of specific casts, I will attempt only to outline the principles that are intrinsic to all casts:

1. The rod, reel, weight, and line must be a matched and harmonious unit; otherwise, the caster's physical effort goes for naught and results in a dismal cast. Trying to cast an eight-ounce sinker with a willowy seven-foot rod is a sure recipe for failure.
2. Timing is more important than strength in casting. While it is certainly true that the angler's physical size confers some advantage with regard to achieving a long cast, most surf casts will be well short of the distances (500 to 600 feet) reached by profes-

sional casters in tournaments. Accelerating the rod, stopping the rod, and releasing the line—all parts of the casting stroke—need to be choreographed together to achieve an accurate cast.

3. To effectively load a long surf rod, first begin with three to four feet of line between the rod tip and the rig and then start the cast slowly. If the cast is initially rushed, the rig will move too fast and prevent the rod from bending properly. Once the rig is moving, accelerate the rod to reach maximum speed just before the end of the casting stoke. At this stage, the rod will be fully loaded.

4. Proper technique calls for pushing the section of the rod above the reel with your right hand and pulling near the butt with the left hand during the acceleration period. (Reverse the hand placement if your left hand is dominant.)

5. Stopping the rod abruptly after acceleration is critical to releasing the energy stored in the loaded rod.

6. The lure or rig will travel in the direction the rod tip moves just before you stop the rod. If the tip points to the right of the intended target, accuracy suffers, your line might cross the lines of fellow anglers, and distance decreases markedly.

7. The casting arc, the distance the rod tip moves during the cast, is short for short casts and long for long casts.

LOCATING FISH:
FEEDERS, SLOUGHS, AND RIP CURRENTS

While many factors make up a successful surf fishing outing, probably none is more important to the angler than the ability to "read the water"—to analyze where and how waves break to determine the structure of the sandy bottom. As waves roll in from deeper water, they will slow down because the shallow bottom exerts a frictional drag on their forward movement. The decrease in wave speed results in a shortening of the wave (wave length decreases) but a vertical growth of the wave (wave height increases). The waves will grow until they become unstable and, like a house of cards, topple over in the form of breakers. Since water depth is the main factor controlling breaker formation, waves break in shallow water and not over holes, depres-

Sandbar and slough (author's photo)

sions, and sloughs. Scanning the water reveals places where waves are breaking in one place but not in another.

If the seafloor within the nearshore zone has a flat, gentle slope, which is representative of many of the beaches along the Outer Banks, the incoming waves break gradually over a wide swath of the nearshore zone. These breakers are known as spillers because they break by spilling water from the crest down the front face of the wave. Being able to recognize these breakers is a clue to the observant angler that the bottom is relatively homogenous and lacks any significant bottom features that may attract both prey and predator alike on a regular basis. This shallow, unremarkable bottom may warrant a couple of casts, but time is better spent searching for more productive sites.

Along stretches of the Outer Banks, the nearshore bottom takes on a more interesting appearance, at least from the perspective of the angler. Sandbars are the most common type of bottom structure found within the nearshore zone. They can stretch for miles along the beach and are evident where the waves crest and break in the form of plunging breakers. These plunging breakers are curling waves that form when the seafloor slope changes rather abruptly. As a result, the crest of the wave travels fast enough to outrun the rest of the wave and plunges downward until it crashes into the trough preceding the wave. Coupled with the sandbar is a slough, often quite deep, which separates the sandbar from the beach. The slough is revealed by the absence of breakers. In essence, waves will break over the bar, reform in the slough, and finally break against the shoreline, dissipating their energy.

Once you can recognize this basic bottom structure, the next step in the process is deciding where to place your offering. Fish such as spotted seatrout and kingfish will generally cruise the slough; they feel comfortable there because the deeper water affords them a certain degree of anonymity in their search for food. Also, small baitfish, the major forage items for many game fish, are often concentrated in the slough, so the optimal location for lobbing a chunk of bait would be into the slough. Some anglers initially place their offering on the sandbar and then let it tumble into the slough. This erratic movement often catches the attention of a fish that is holding near the bottom and waiting for food items to be washed down to it by the breaking waves.

If there is an opening in the bar, as evidenced by the lack of breaker activity, this can be a prime site to fish. The breach provides a conduit for both game fish and bait to move inside the bar. A well-placed cast near this opening will often intercept a variety of predators, both small and large, as they search out their prey.

There are stretches within the seashore that are marked by two sandbars and two sloughs that run parallel to the beach. In this situation, two sets of breakers will be visible over the sandbars; the outer sandbar will have larger breakers than the inner sandbar, and the sloughs will have little or no breaker activity. As with the above case, the sloughs often prove to be productive, but the outer slough will entail a longer cast that may be beyond the ability of some anglers.

For the novice surf angler, scanning the nearshore region during low tide rather than high tide may prove more productive in locating sandbars and sloughs. When there is a minimum amount of water covering the bottom, all of the structure may be visible, and subtle variations in bottom topography, which may hold fish, become evident. The downside is that during very low tides the water depth in the slough may be too shallow to attract fish, and the angler should return at high tide.

Another surf structure that is attractive to game fish is an "out-suck," more appropriately known as a rip current—a small-scale nearshore current

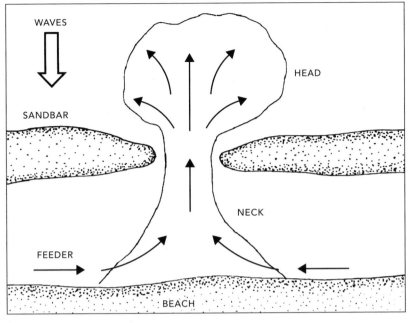

WAVES

SANDBAR

HEAD

NECK

FEEDER

BEACH

Rip current

moving away from the beach at a relatively high speed. Rip currents form as water in the surf zone is continuously transported to the beach. In the long run, the water cannot simply accumulate on the beach, but it must flow back through the surf zone in narrow corridors. The return flow through these corridors consists of all the water that was transported toward the shore over a broader width of beach. A rip current has three parts: feeder currents that flow parallel to the beach, the neck in which the feeder currents converge and flow through the breakers in a narrow corridor, and the head in which the current weakens and widens beyond the breaker line in deeper water. Anglers can recognize rip currents by looking for these clues: a channel of choppy water; an area having a notable difference in water color or turbidity; a band of foam, seaweed, or debris moving steadily away from the beach; and a break or disruption in the incoming wave pattern. From beach level, rip currents may be difficult to see; anglers will obtain a better view of the water conditions by climbing a dune or standing on their ORV. Rip currents can be particularly strong during storm conditions and may also increase in strength during low tide. Anglers, particularly those burdened with heavy tackle and waders, need to be cognizant of the danger of wading near the strong outflow.

Rip currents can be magnets for game fish like striped bass and red drum, which are comfortable in moving, swirling water. Their broad tails allow them to efficiently navigate the rip current and ambush small prey that are swept away from the inner slough by the strong flow. These species will often congregate around the head of the rip current where there are deeper holes, but under the cover of darkness, they may move into the neck and even into the feeder zones. Anglers should attempt to cover all the parts of the rip current with their casts; however, if the head extends beyond the outer sandbar, long casts are needed to reach this target zone.

You should be aware that this stretch of the North Carolina coast is quite dynamic; the shoreline and the nearshore zone change frequently due to seasonal variations in wave size and direction, currents, and storms. A section of the beach that was productive during one time of the year because of its unique structure might have changed considerably over time and may now hold few fish. It is incumbent upon the angler to search out new "fishing holes." While surf anglers, as a lot, are often very helpful and willing to share advice and techniques, they often guard their "secret spots" with utmost care. But there is nothing secret about the following sites, which receive their fair share of angling pressure.

Cape Point

During the spring and fall, prime surf fishing times, Cape Point, or simply the "point," may have scores of vehicles and anglers at its very tip. These anglers will often stand shoulder to shoulder in knee- or chest-deep water for hours at a time. If you desire solitude and space, then the Cape Point scene is not for you.

Ask anglers why Cape Point is considered the epicenter of surf fishing along the Atlantic coast, and myriad responses come forth. But the most prevalent reply is that Cape Point and its submerged extension, Diamond Shoals, is where the cold, southward-flowing Labrador Current collides with the warm, northward-flowing Gulf Stream, setting up a dynamic interplay between contrasting water types that attract both northern and tropical game fish species. The argument is flawed because though the Labrador and Gulf Stream currents do indeed meet, their confluence is well offshore of Cape Point. In particular, the Gulf Stream is a deep-flowing current that extends hundreds of feet below the sea surface, and this characteristic limits

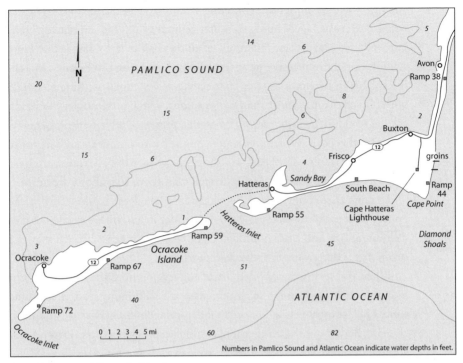

Cape Hatteras and Ocracoke Island

it to the deeper reaches of the Atlantic basin and not to the shallow near-shore zone.

While it is true there are often pronounced temperature differences between the northern and southern waters of Cape Point, these thermal gradients are the result of local nearshore currents and, at times, the upwelling of cooler bottom water north of Cape Point. To explain the attraction of Cape Point for game fish, we fall back on a fundamental ecological principle, albeit simply expressed: organisms go where there is food. As deepwater waves approach the shallows of Diamond Shoals, they often bend or refract toward the shallower water, converge, and stir up nutrients from the bottom, initiating a food chain with game fish at the top.

Ask any hard-core surf angler to articulate Cape Point's attraction, and two words are generally spoken: red drum. At times, massive schools of these fish congregate at this site, their movement triggered by the presence of one of their favorite baitfish: the finger mullet. Beginning in September, the first wave of mullet starts leaving Pamlico Sound, and this migration peaks in October. Depending on water temperature, this "mullet run" may last well into November. The result of all this bait in the water is that large schools of red drum converge at Cape Point to feast on this bounty. Matching the hatch becomes the key to success for any angler wishing to tangle with big red drum of more than forty pounds. While many anglers will use a castnet to catch their own supply of mullet, most will rely on the local tackle shops for their bait. Whichever route you choose to procure your bait, make sure the bait is fresh. Big drum will simply not touch mushy, timeworn bait. Change the bait frequently; every fifteen minutes or so is about the norm.

To tame these bullish drum, most surf anglers come equipped with the appropriate tackle. Stout surf rods, twelve feet in length, are needed to toss a chunk of bait and a heavy sinker (eight or more ounces), as well as play a drum that is determined to seek its freedom.

When the surf runs red with drum converging on Cape Point, it is not uncommon to find scores of anglers who are all flinging lead at the same time. This "combat fishing" requires skills and techniques that are unique to this site. Because of the congestion, casting accuracy is a top priority—cast straight out from your spot to avoid crossing and tangling other lines. Monitor your line religiously. If the current is running strong, as it often does at Cape Point, your rig will tumble along the bottom. To avoid the inevitable entanglements that will result, walk down the beach with the current and keep parallel with your line and rig. Once you are in slack water, reel in, walk back up the beach, and enter the "conga line" again.

If you're fishing but not catching, while those around you are hooking up, a longer cast may be in order. Drum are prone to hang around the drop-offs where the water is breaking over the shoal. During low tide, this deeper water may be out of the reach of an anemic caster.

There are no guarantees when fishing the point; one may see a dozen fish or a hundred fish in a bite. But many drum anglers swear by the following maxim:

Wind northeast,
Fish bite least,
Wind southwest,
Fish bite best.

Though red drum are the signature species at the point, cobia make their appearance from about mid-May to the end of June, and small bluefish and

Spanish mackerel are prevalent throughout the summer. Striped bass show up with the cooling waters of winter.

Outer Banks Inlets

After Cape Point, the inlets—Oregon, Hatteras, Ocracoke, and Drum—attract their fair share of angling activity. The existence of tidal currents—the horizontal movement of water due to the rise and fall of the water level—is a distinguishing feature of these inlets. The positions of the moon and the sun in relation to the earth control the amount of water that flushes into an inlet, a flood tidal current, and out of an inlet, an ebb tidal current. These orientations, which can vary on both a daily and a monthly cycle, affect not only the timing but the speed of the currents.

During a twenty-four-hour period, the Outer Banks generally experience two high tides and two low tides that have a tidal period (the elapsed time between successive high or low tides) of approximately twelve hours. The rate at which the water level rises and falls during a tidal cycle as well as the speed of the current is not constant; it gradually changes at the beginning and end of the cycle and peaks in the middle of the cycle. Current speed also increases when water flows through a narrow opening, such as an inlet or sandbar. Conversely, a strong onshore wind may significantly reduce the strength of an ebb current and result in only a small volume of water leaving the estuary.

The feeding habits of many fish are directly affected by these tidal movements. With an ebb current, fish will often stack up at the mouth of the inlet to intercept the bait that is washing out from the sound and time their feeding activity to current strength. In general, many fish are more aggressive feeders during the weaker stages of an outgoing tide and may stop feeding during peak flow to avoid having to fight the full strength of the current. And yet if fish can relocate to a sheltered spot, such as a deep drop-off, they can lie in wait to ambush prey that are trapped in eddies that spin off from the current.

Before the tide reverses, a short period of slack water occurs. As ocean water begins to flow into the sound, fish adjust with the changing flow and move with the current to target food items riding in with the tide.

A number of years ago, while fishing the south side of Oregon Inlet during late November, I experienced firsthand the influence of tides on a fish's feeding cycle. A strong flood current was streaming through the passage—

so strong that my fellow anglers had to follow their lines down the beach as their sinkers tumbled along the bottom. Big bluefish, ten- to twelve-pounders, were stacked up in the inlet to intercept the drifting offerings. These fish didn't blitz the beach, ignored surface plugs or lures, and were content just to hold near the bottom. Bluefish after bluefish were caught during this period of inflowing water, but the catching came to an abrupt halt when the flooding current weakened. The blues were gone; my casts simply fell upon barren waters. I returned to the inlet the next day, not knowing what to expect, but like workers punching a time clock, the blues swarmed in with the high tide.

If the moon were stationary, the timing of the tides would not change significantly from day to day, but the moon revolves around the earth. In one day the moon travels approximately thirteen degrees of a circle and appears over the same location on earth approximately fifty minutes later each day. For the angler, this astronomical interplay means that a specific tidal stage is fifty minutes later each day. (If the first high tide of the day occurred at 0600, on the next day it would peak at 0650.)

Twice a month, during the new moon and full moon phases, the moon and the sun are in alignment with the earth, and this celestial orientation results in spring tides—the time of greatest variation between high and low tides. Correspondingly, during the first-quarter and third-quarter phases, the moon, sun, and earth form a right angle, and this alignment marks the period of neap tides—the smallest tidal range during the month. All other factors being equal, the strongest tidal currents will occur during the spring tides and the weakest during the neap tides. The large volume of water that pours though an inlet during the spring tides may present a challenge to the angler because it spreads out the area where fish may be holding.

Fortunately, the average angler doesn't have to be well-versed in celestial mechanics to predict the tides. Tidal forecasts are made from relatively long-term tide observations that are recorded continuously at many sites around the globe. In essence, the study of past tidal records yields future tidal times that are available in the form of a tide table from local tackle shops. This information is invaluable if you want to fish a particular tidal cycle.

Cape Hatteras Lighthouse

The lighthouse's groins (locally, they are often called jetties) can often hold fish. Though in various stages of decay, these man-made structures project into the ocean from the beach. The groins were constructed with the intent of widening the beach in front of the lighthouse before the National Park Service (NPS) decided to relocate it. Since the lighthouse was built in 1870 (for the second time), it has seen an ever-encroaching sea. (The lighthouse originally stood 1,500 feet from the water.) On a yearly average, coastal currents carry their load of sediment (littoral transport) from north to south. The groins block this transport of sediment, and sand piles up on the upstream side of the groin. However, on the downstream side of the groin, lacking its natural source of sand, the beach narrows and the water deepens. Since numerous fish species have an affinity for hard structure because it provides habitat and bait, knowledgeable anglers will make the trek over the dunes to the groins and target spotted seatrout and flounder.

HANDLING FISH:
THE ETHICAL ANGLER

You are probably reading this book because you have a keen interest in angling for the fish that swim the waters of the Outer Banks. With the growth in the popularity of Outer Banks fishing, particularly surf fishing, must also come a new breed of educated and concerned anglers attuned to the well-being of the species they eagerly pursue, without which there can be no quality fishing.

One popular outgrowth of this movement toward a more knowledgeable angler is the practice of catch-and-release. Though Michigan fly anglers instituted this practice during the 1950s as a means of reducing the expense of stocking trout streams, its current acceptance is widespread throughout the angling world. An angler may release a fish for a variety of reasons—wrong species, wrong size, and conservation issues—but implicit in the release of the fish is the angler's belief that the fish will survive. Limited studies have shown that most fish released after capture survive, but the survival rate is highly dependent on the actions of the angler.

You've caught that fish of a lifetime and you're going to release it, but how should the fish be handled? If possible, keep the fish in the water, minimize contact, and use a dehooking tool to safely and easily remove the hook. If the fish is taken out of the water, as in most surf fishing cases, minimize its exposure to air. With prolonged exposure due to removal of hooks, measuring, or photographing, gills collapse, with the individual filaments adhering to each other, which can lead to a rapid decline in blood oxygen levels, tissue damage, and eventually death.

While the fish is out of the water, support its weight horizontally and avoid holding the fish, especially a large fish, by its jaw. A fish held for a prolonged period in the vertical position can experience damage to its internal organs.

Finally, if your fish is in good shape, put it back into the water headfirst. Note how the fish responds to being back in its environment. If it is lethargic or exhibits erratic behavior, the fish may require some additional help before it can swim away on its own. You can revive a moderately exhausted fish by first placing one hand under the tail, holding the bottom lip (be careful with toothed fish) with the other, and gently moving the fish forward. At the first sign of the fish's revival, let it go—repeated attempts at resuscitation will only induce more stress on the fish.

While your intentions to release a deep-hooked fish might be admirable, this fish might be a good candidate to keep because its chance of survival is low. If the fish must be released due to size or bag limit regulations, then the odds of survival can be increased if you do not attempt to remove the hook. Simply cut the line as close to the hook as possible and release the fish. Remarkably, some fish that are hooked in this manner are able to expel the hook from their mouths within a few days.

THREATS TO SURF FISHING

A battle over a small, beach-dwelling bird has pitted environmentalists against a legion of beach users. The bird is the piping plover that was abundant in the nineteenth century but was nearly wiped out due to hunting for its plumage. Now protected by the Endangered Species Act of 1973, the piping plover has gained support from the U.S. Fish and Wildlife Service, which wants to designate the seashore as critical habitat for the bird and thus limit and, in places, restrict beach access. The nesting preferences of the piping plover are at the crux of the debate: accreting ends of barrier islands, sandy peninsulas, and tidal inlets—all prime angling sites within the seashore. Environmental organizations like Defenders of Wildlife argue that piping plovers are particularly vulnerable to vehicular traffic because they feed primarily during daylight hours and search for prey at or near the sand-water interface or in the flotsam that has washed ashore. Counterarguments from the Outer Banks Preservation Association, a grassroots organization on Hatteras Island, point out that the seashore is at the southern end of the migratory path of the piping plover, and the negative economic impacts of beach closures could be catastrophic. As of this writing, this contentious issue resides essentially in the hands of the NPS, which has a federal mandate (in place since the 1970s) to develop a comprehensive ORV plan that would provide the necessary management and regulatory framework for continued ORV use within the boundaries of the park.

Currently, the beaches are being managed under the terms of a consent decree ordered by U.S. District Court. According to its terms, resource closures go into effect during the birding season from mid-March to approximately mid-August.

Pier Fishing

Many of the same surf species, such as flounder, kingfish, and bluefish, are readily available to those who walk the wooden planks, but as we will soon see, fishing techniques may be quite different. The obvious advantage that a pier angler has over a shore-bound colleague is the ability to cover more water. Since the Outer Banks piers extend hundreds of feet into the Atlantic, pier anglers can target species that normally do not enter the nearshore zone. In addition, long casts are often not needed, since anglers can simply plunk their offering down into the slough or some other structure that the pier spans.

The piers are open seasonally from April to December and remain accessible well into the night. Each pier has a pier house where anglers can purchase bait and tackle as well as rent rods and reels. Though there is a fee (as of this writing, a daily pass is $10), pier anglers do not need a saltwater license.

Pier fishing has been popular in the Outer Banks for more than seventy years. The original Jennette's Pier in Nags Head is said to be the oldest, tracing its beginnings back to 1939. Though this venerable wooden structure had survived numerous hurricanes and nor'easters, hurricane Isabel in 2003 destroyed 540 feet of it. In 2007 the state acquired the pier, demolished it, and set into motion a plan to rebuild it. The new pier will extend 1,000 feet into the ocean, will be designed to withstand 130-mile-per-hour winds, and is expected to be completed by May 2011. Unfortunately, the Cape Hatteras Fishing Pier in Frisco has not fared as well. It has been damaged by storms, and there are no immediate plans to reopen it. (Hurricane Earl, which brushed the North Carolina coast in September 2010, caused additional damage, breaking the pier into four sections.)

Is one pier better for fishing than another? I can't arrive at an unequivocal answer because the appeal of pier fishing is different for each angler: some want a variety of fish, and others want to catch their limit. But it appears that fishing reports consistently cite Avon Fishing Pier for its red drum fishing during the fall. The world-record red drum was caught about 200 yards from the pier in 1984.

Pier fishing (author's photo)

BELOW THE PIER

A pier allows you to obtain a bird's-eye view of the water that is not possible from the shore. While this expanse has its own intrinsic beauty, a more critical study of the happenings below can increase angling success.

In particular, note the direction in which waves travel as they progress toward the shore. Rarely are these waves parallel to the shoreline, but rather, the crest lines of the waves are at some angle to the shore. (An offshore storm north of Nags Heads Pier will generate waves that move from the northeast and strike the essentially north-south-oriented shoreline at an angle.) Though these waves will refract or change direction slightly as they enter the shallow water, when these waves strike a beach and release energy, they generate currents, known as longshore currents, in the surf parallel to the shoreline. The speed of the current is highest in the mid-surf region and increases as wave height and angle of wave approach increase. A storm out at sea can generate large offshore waves that roll shoreward and set in motion a strong longshore current. Under these conditions, even a six- to eight-ounce sinker will not hold bottom but tumble down the beach.

The flow of the longshore current can shift back and forth as the direction of wave approach changes from day to day. As you face the ocean, the longshore current may be moving either north or south along the shoreline. However, the predominant current direction along the Outer Banks is to the south. Since the current sweeps food down the beach with it, predators may lie in ambush near the pier's pilings, waiting to intercept their next meal. In addition, the sands of beaches and nearshore zones are continually being moved by longshore currents. Too much sediment roils the water, decreases light penetration, and hinders predators that are sight dependent. If the longshore current is quite strong, fish may seek shelter on the down-current side of the pier. You should adjust accordingly and target bottom-feeding species (spot, croaker, and kingfish) that strongly depend on their sense of smell to locate food. (See Chapter 8 for a discussion of the anatomy of a fish as it pertains to its sense of smell.)

Farther out on the pier the water below may appear at first glance to be nondescript. But when viewed with quality polarized sunglasses, which significantly eliminate glare, variations in the bottom topography become evident. Depressions and holes, though small in size, may hold fish. Also, tiny baitfish that congregate around pilings are much more visible.

Similar to surf fishermen, pier anglers can spot the location of the near-shore slough by the absence of waves. But if the winds are calm and wave activity is at a minimum, how can you pinpoint the slough? Look for subtle variations in water color. The color of the water over the slough will be a little more vivid than it is over the shallower bottom due to the selective absorption of certain wavelengths of light with depth.

With waves breaking over it, particularly at low tide, the nearshore bar is readily evident even to the most casual observer. The turbulent, shallow water that surges over the bar offers little advantage to foraging fish or fish seeking shelter. Thus the site will generally not hold a significant number of fish, and anglers should seek out more promising locations. Fortunately most of the piers extend beyond the bar out into the deeper water of the Atlantic. While depths don't increase markedly, anglers can expect to encounter depths of more than thirty feet, which can attract species, such as king mackerel, that prefer this deeper and generally clearer water.

The pier is a magnet for a variety of game fish that depend on organisms lower on the food chain. These include bottom-dwelling animals, such as barnacles and mussels, that attach themselves to pilings. These filter feeders, which sieve the water for planktonic organisms, are a favorite of sheepshead. Small baitfish, such as silversides and anchovies (glass minnows), will seek shelter and refuge around the pier in an attempt to avoid predation by speedy predators, such as Spanish mackerel.

ENVIRONMENTAL FACTORS AFFECTING FISHING: TEMPERATURE AND WINDS

Most marine species are comfortable within a well-defined temperature range. If they are out of their comfort zone, their only option is to seek waters that satisfy their temperature needs. Or, viewed another way, because these organisms have thermal limits, geographic limits are fixed latitudinally, and the range of many Atlantic species coincides with Cape Hatteras, which experiences a major latitudinal variation in temperature. For example, pier anglers on the northern Outer Banks should not expect tarpon to routinely travel into these waters. The North Carolina state-record tarpon, weighing over 193 pounds, was caught in 2008 from Sea Fishing Pier in North Topsail Beach, which is south of the Outer Banks.

Migration patterns are often tied to seasonal temperature changes. As water temperatures cool during the fall, small bluefish move southward from their summer feeding grounds. But want that citation-sized bluefish?

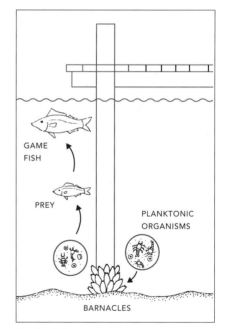

Pier food chain

You have to wait as temperatures drop even lower during late November and early December, as the partition of bluefish by size is in part controlled by differences in water temperature preference between juvenile and adult bluefish. The larger fish appear later in the year than the smaller ones.

Fish, in general, are poikilotherms — organisms whose body temperature corresponds to that of the environment. Since these species are unable to regulate their internal temperature, they are susceptible to rapid changes in water temperature, such as that which results with the passage of a strong cold front or Arctic outbreak. Even a few degrees' change in temperature may not cause them to leave the area but might alter their feeding habits and patterns.

Temperature Tolerance of Selected Game Fish

Species	Lower limit(°F)	Optimal(°F)	Upper limit(°F)
Amberjack	60	65–75	80+
Bass, Striped	50	55–75	75
Bluefish	50	66–72	83
Dolphin	70	72–78	84
Drum, Red	52	70–85	90
Flounder (summer)	56	62–66	72
Jack, Crevalle	65	70–85	90
Mackerel, King	65	68–76	88
Mackerel, Spanish	64	70–81	90
Marlin, Blue	70	74–82	88
Marlin, White	65	68–78	82
Pompano	65	70–82	85+
Sailfish	68	72–82	88
Seatrout, Spotted	48	68–78	87
Snapper, Red	50	55–65	70+
Tarpon	70	75–90	95+
Tuna, Bluefin	50	60–72	82
Tuna, Yellowfin	64	72–82	81
Weakfish	45	55–68	78

Most pier houses will post updated daily water temperatures—information almost as valuable to the angler as what is biting.

Wind direction can play a critical role in pier fishing. During the summertime, winds blow predominantly from the southwest. These winds tend to stir up the nearshore bottom sediments, significantly decreasing water clarity. Plumes of sediment-laden water can be seen extending from the shoreline seaward. Sight-dependent species, like Spanish mackerel, will assiduously avoid this "dirty" water. If possible, move down to the end of the pier, where, in general, the water will be clearer and less turbid. If these winds persist for an extended period, the cloudy water will reach beyond the end of the pier and most likely cause these species to stay father offshore.

In contrast, northeast winds will usher in clearer ocean water and, with it, the appearance of pelagic species, such as king mackerel.

Southwest winds can also significantly change the water temperature and the composition of species that inhabit the nearshore zone. Due to the complex interaction of the wind's drag on the water surface and the earth's rotation, surface water is transported ninety degrees to the right of the wind, or in the case of a southwesterly wind, water is transported away from the shoreline. This displaced water is replaced by deeper, colder water in a process known as upwelling. In the summertime, waters north of Cape Hatteras can be considerably cooler than waters to the south under a regime of strong and persistent southwest winds.

TARGETING FISH: TACKLE AND TECHNIQUES

Angling opportunities on a pier can include fishing for bottom species with a baited rig, casting lures, and live-bait fishing for apex predators. Many anglers who spend a day on a pier may engage in more than one of these activities, depending on what the sea offers up on a given day. However, many pier anglers target small bottom-feeders, "pan fish," and for good reasons: only simple tackle is needed; minimum skill level is required on the part of the angler; and these critters are simply tasty to eat.

A seven-foot rod, which is matched with a medium-sized spinning reel (ten- to twelve-pound test line), will handle most bottom fish. The rod tip should be relatively stiff, capable of lifting fish up and over the railing of the pier. Use enough weight, two to four ounces, to hold a two-hook rig on the bottom but not to overload the rod. While many anglers use squid for cut bait because of its toughness and relatively low cost, pier aficionados have on hand bloodworms and shrimp when the squid is not producing. Bloodworms, though expensive, are a favorite of spot, and shrimp yield catches of kingfish and pompano.

Bluefish, which are available from spring to fall, are caught by pier anglers commonly using two-hook fireball rigs—brightly colored (red or yellow) foam balls in front of each hook. The rigs, which are fished on the bottom, are said to attract the attention of the blues due to their coloration and movement and can be particularly effective when water is discolored. Since

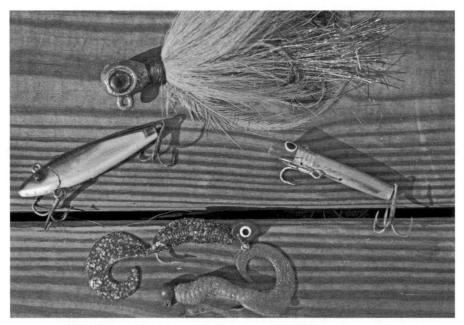

Lures (clockwise from top): bucktail, Got-cha plug, leadhead jigs, and MirrOlure (author's photo)

bluefish use their keen sense of smell to locate food, use the freshest strip or chunk of oily bait, such as menhaden or mullet.

Success in bottom-fishing often comes to those anglers willing to stay mobile. If one section of the pier is unproductive, you should move from shallow water to deeper water or vice-versa—don't hesitate to experiment. Location is the key to pier fishing.

When bluefish and Spanish mackerel are actively chasing bait, anglers switch from bottom rigs to lures. The lures of choice for many pier anglers are Got-cha plugs. These elongated lures, approximately the size of a cigarette, come in a variety of colors, weigh about one ounce, and when retrieved, mimic fleeing baitfish. You can employ the same light rod and spinning-reel combo that you used in the surf. Small schools of bluefish and Spanish mackerel can show up almost anywhere along the pier but are generally not spread out. This concentration of fish means fishing in close quarters with other eager anglers. Casting accuracy should be your first priority. Crossed lines mean time lost, and these fish often move through the area quickly. My technique is to bend over the top rail and cast underhand. While flipping the plug in this manner does sacrifice some distance, it has the advantage of the line traveling straight out from the pier. In addition, I am in position to the see the plug and start reeling.

Manipulating the plug properly is the key to success, and with a little practice, almost any angler can master the technique. Once the plug hits the water and sinks for a moment, the angler simultaneously reels and twitches the rod tip, which causes the plug to dart back and forth just below the surface. The angler should carefully watch the movement of plug upon retrieval to ensure that it is moving back and forth—the "action" that the plug should exhibit. Retrieval speed and manipulation of the rod tip will vary depending on water conditions and the species being targeted. In general, Spanish mackerel prefer a fast retrieval and a lively action to the plug. Work the plug all the way back to the pier, since many strikes occur near the pilings.

Upon hookup, be prepared to control the fish by reeling steadily and maintaining constant pressure. When the fish is near the pier, you should point the rod tip straight down and start cranking the fish up. This technique works particularly well for small- to medium-sized fish, eliminating the rod bend that results as the fish is dangling in midair between the water and the pier's rail.

Got-cha plugs contain two treble hooks—one in the middle of the lure and one at the back. If you plan to release your catch, mash the hook's barbs down and go barbless. Studies have shown that barbless hooks allow for a more efficient release of the fish, produce less tissue damage, and minimize mortality. I have often heard anglers complain that they are more likely to lose fish without the purchase provided by the barbs. Not so. Maintaining pressure on the fish—a tight line—leads to success.

While bluefish and Spanish mackerel may show up during any time of the day, many pier anglers concentrate their efforts during the early morning and late evening hours—prime times for these species. Bluefish, in particular, feed very little during the night, content to conserve their energy resources for hunting during the day. But by sunrise they are famished, and the primal need to feed propels them into high gear. Their swimming speeds increase markedly just before dawn, and when the first glimmer of sunlight breaches the horizon, they are roving the water column in a frantic search for their first meal of the day. I recommend working the length of the pier, since these fish can be anywhere there is a concentration of bait.

Both species travel in schools, which may be intermingled, but since bluefish can tolerate lower water temperatures than Spanish mackerel, they usually make their presence known to pier anglers a few weeks before the arrival of the first mackerel. Spanish mackerel start to show up when the temperature reaches 70°F and become numerous with temperatures approaching 72° to 75°F.

For a number of years, I have used a tandem bucktail rig, which consists of a bean-shaped bucktail (one and a half ounces) and a trailing teaser (one-eighth ounce). The rig can be fished to imitate swimming baitfish or jigged vertically through the water column. With regard to the former, the retrieve is similar to fishing a Got-cha plug, but slower and more deliberate. I particularly like to fish the surf zone, so that the longshore current can sweep the rig downstream to waiting fish. Though I have caught both bluefish and mackerel on this rig, weakfish are my targets, and they will eagerly strike this rig. (Because bluefish and mackerel have sharp teeth, I don't recommend using this bucktail combination as your main lure.) Bouncing the rig off the bottom when retrieving often catches the attention of sedentary, ambush predators, like flounder. If the pier is not crowded, try casting the rig parallel to the pilings, working the bucktails throughout the water column. In this manner, you can cover more water that is near prime structure.

Most piers have a T-shaped end section that is generally used by hard-core "plankers" who have their sights set on big fish, like king mackerel (see Chapter 9 for more information on this fish). This type of angling requires two rods. One rod serves as an "anchor rod" that is employed to cast out a heavy, multipronged sinker; its sole purpose is to securely grab the bottom. The anchor rod is generally a stout surf rod (ten to twelve feet) capable of throwing the weight a long distance. The "fighting rod" is a thirty- to fifty-pound class boat rod (six to seven feet) and is matched with a similar class conventional reel. On the fighting rod, live bait is impaled on a hook, and the whole rig slides down the line of the anchor rod via a clip that is initially attached to both lines. The bait is generally hooked under the dorsal fin, and the bait should swim just below the surface. When the fish strikes the bait, the clip pops loose from the anchor line, allowing the angler to fight the fish.

Live-bait fishing is a waiting game; patience is a definite plus, as hours may pass without even one strike. During that time, both rods are placed against the pier railing, and the reel's drag on the fighting rod is set lightly, so that the rod is not pulled over when there is a hookup with a large and determined fish. In the interim, anglers may try to restock their supply of live bait (spot, pinfish, and small bluefish) and keep them alive in a bait bucket that is suspended in the water under the pier until needed. And they will be needed. During the course of a day, anglers may use more than a half-dozen baits to replace those that have become fatigued or been preyed upon by larger species. Frisky bait is always a plus.

When a fish strikes the bait, essentially hooking itself, the idea is to let the fish run with the bait. Even with the light drag setting, the friction of the line moving through the water will exert enough drag to slow the fish down, hopefully. The angler then plays the fish back to the pier, where teamwork is necessary for the angler to retrieve the prize. Fellow anglers can help to untangle lines that often become crossed, but what is needed most is someone to lower a net to hoist the fish up to the pier. Hefty fish in the thirty- to forty-pound range cannot be reeled up to the pier's railings.

News spreads fast when a trophy fish is landed, and the end of the pier can become quite congested, with rods covering almost every available space along the railings. These crowded quarters should be assiduously avoided by anglers who want to bottom-fish or cast lures, unless they desire hardened stares or choice words from the live-baiters.

Though many of the above fish are available from spring to fall, when the water cools, many of the smaller fish migrate southward, while others move offshore to waters that stay warm enough for them to make it through winter. Spot, croaker, flounder, and some bluefish winter over in North Carolina's offshore waters, so many anglers turn their attention to red drum. By late October, these are big fish. Coupled with rough surf that is common at this time, they require the use of stout tackle. Rods capable of tossing eight- to twelve-ounce sinkers plus a chunk of fresh bait on a fish-finder rig from the end of the pier are needed to reach the deeper water that may hold drum.

The last chance for pier anglers to score comes with the arrival of striped bass during late November and December. Though their numbers have been fairly constant over the last few decades, this was not always the case with the striped bass population. Records show that from the 1960s to the mid-1980s, striped bass infrequently visited the Outer Banks, resulting in small catches. Conservation efforts in the 1980s, centered on the spawning population in the Chesapeake Bay and the Roanoke, Delaware, and Hudson Rivers, were successful in restoring the fishery. Today, anglers mainly fishing the piers north of Oregon Inlet can catch their limit using medium-sized (fifteen- to seventeen-pound) spinning tackle to cast lures or bait.

THREATS TO PIER FISHING

Fishing from the piers in North Carolina may be a dying pastime; both Mother Nature and development appear to be conspiring to close down the piers. Over the years, nor'easters and tropical storms have ravaged the piers, destroying some and severely damaging many others. As we have seen, piers along the Outer Banks have not been spared. While repairs have been made, their cost has spiraled, and owners are reluctant to rebuild again with the threat of more storms looming in the future. At one time, there were eight operational piers along the Outer Banks, but only five are currently open. (Recall from Chapter 1 that I, as many do, mark the southern boundary of the Outer Banks to be Beaufort Inlet. But there are those who refer to Bogue Banks, which stretches from Atlantic Beach to Emerald Isle, as part of the "southern" Outer Banks. This part of the coast has a number of piers.)

Coastal property is probably the most expensive real estate in the United States. With the coastal communities already packed with houses, stores, and shops, developers are exerting considerable pressure and offering lucrative financial packages to convert the seafront property on which the piers sit into high-end homes.

Sound Fishing

There are seven sounds—Currituck, Albemarle, Roanoke, Croatan, Pamlico, Core, and Back—that comprise the backwaters of the Outer Banks and are known to coastal geologists as lagoons, shallow estuaries that are separated from the ocean by the barrier islands or barrier spits. These semi-enclosed bodies of water have connections to the open ocean via inlets, and within the sounds seawater is measurably diluted by fresh water derived from land drainage. This mixture of fresh and salt water results in a brackish concoction. Within this environment of intermediate salinity, menhaden, mullet, and shrimp thrive, to be ultimately preyed upon by top-level predators in this estuarine food chain.

But the salinity of some of these sounds has changed rather markedly over time. Currituck Sound was once viewed as more of a big freshwater lake than a coastal estuary, populated by freshwater grasses, like wild celery and pondweed, and freshwater fish, such as largemouth bass. This low salinity was the result of the closure of Currituck Inlet in 1828, which eliminated the link between Currituck Sound and the ocean, and the input of fresh water into the sound from rivers (North Landing River and Northwest River) draining the Virginia and North Carolina mainland. Currently, the salinity of Currituck Sound is higher than it has been for decades, mainly due to the input of brackish/salt water from a canal in Virginia. This increased salt content has led to a decline in the freshwater-loving largemouth bass.

In contrast, the salinity of Pamlico Sound can vary markedly within the sound itself. The Neuse and Pamlico Rivers flow in from the west, and the salinity can be quite low (five to fifteen parts per thousand) at the mouths of these rivers. But twenty-five miles to the east, near Ocracoke and Hatteras Inlets, where tidal currents flood through these narrow gaps in the barrier islands, the estuary is infused with seawater, and salinity values are approximately that of the ocean (thirty-five parts per thousand).

On the eastern seaboard, Pamlico Sound is the largest lagoon, stretching ninety miles from north to south and twenty-five miles across at its widest point, from Cape Hatteras westward to the North Carolina mainland. In contrast, Core Sound is considerably smaller than its northern counterpart,

Core Sound salt marsh (author's photo)

comprising 88,000 acres, compared with 128,000 acres for Pamlico Sound. The remaining sounds are even smaller.

The biological vitality of the sounds rivals that of their arguably better-known sister estuary, the Chesapeake Bay. The sounds not only support a robust recreational fishery but are also important nurseries for 90 percent of the commercial species caught in North Carolina.

Key components in these vigorous ecosystems are the salt marshes and sea grasses that rim the sounds' shorelines. In their beginnings, the sounds were relatively sterile, lacking the lush, vegetative mats of today, but over time the barrier islands trapped oceanic sediments, forming large overwash fans. Storm surges, which routinely impact the low-lying barrier islands, often carry sediment-laden water over the dunes to the sounds. In these relatively quiet waters, varieties of salt-tolerant grasses take root in the soft substrate and colonize the sediment fans.

Within the sounds' intertidal zone, the signature marsh species is smooth cordgrass (*Spartina alterniflora*), a plant that is responsible for much of the sounds' biological productivity. *Spartina*'s narrow, tough blades and unique glands that secrete excess salt make it ideal to withstand high temperatures and long periods of immersion in salt water. Over time, *Spartina* develops an extensive root system that firmly anchors it in the soft sediments. The plant's tough stalks appear to prevent much grazing by birds or mammals, and usually much less than 10 percent is consumed by the grazers. The bulk of the grass decomposes and, in the process, becomes the dominant food source for myriad low-level consumers that form the links in a fish's food chain. In spring and summer these plants are lush and green, but by fall *Spartina* begins to turn brown as blades die and decomposition occurs. Much of this detrital matter (organic material resulting from the disintegration of dead remains) is flushed out of the marsh and carried to the open water by the falling tide.

In the soft bottom sediments of the sounds, broad flats stretching from the low-water mark to a depth of about five feet are inhabited by sea grasses. These shallow, subtidal, sea-grass beds are composed mainly of eelgrass (*Zostera marina*). Eelgrass is neither a grass nor seaweed. It is an angiosperm, a flowering plant that can live for many years. The wading angler can easily recognize eelgrass by its long, narrow, and ribbonlike leaves, which gently sway with each passing wave. Like *Spartina* marshes, eelgrass beds are characterized by a large complex of rhizomes, a root network within the sediment. The density of the root system, coupled with the thickness of the eelgrass beds themselves, provides a refuge for juvenile fish, crabs, and other animals from the probing eyes of larger species intent on finding their next meal. (Don't look for large flounder within these beds, as these flat fish have trouble entering this estuarine jungle.)

All of the sounds are also noted for their vast expanse of shallow water; maximum depths approach twenty feet. While the lack of deep water makes the area hazardous for large vessels, the shallow water is ideal for wading or kayaking to your favorite fishing spot. In addition, the shallow nature of the sounds makes them more susceptible to wind-driven tides than to astronomical tides, except in the vicinity of their inlets. This phenomenon is particularly true for Pamlico Sound, where the lunar tide is only on the order of four inches. A strong wind from the southwest,

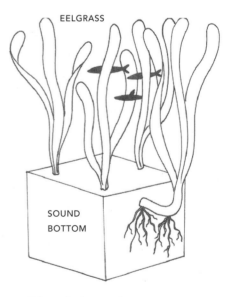

Eelgrass in the sound

which is the persistent direction during the summer, produces a high-wind tide in the northern part of the sound. Conversely, a northeasterly wind displaces sound water to the southwest and results in a similar wind tide in the southern end of the sound. Under these conditions, the angler must realize that tidal tables showing the times of astronomical high and low tides may be of little value when fishing the sound. With the northeasterly wind scenario, the water level in the vicinity of Oregon Inlet is lowered, and the incoming phase of the tidal cycle at the inlet is prolonged; thus a larger volume of water is introduced to the sound through the inlet. A hard southwest wind produces the opposite reaction; the inlet will constantly ebb as water levels increase in the northern sound. The combination of wind tides and shallow depths also results in an unstable bottom; sand shoals are constantly changing their shape and size as well as their position.

There are seasons when fishing success depends more on the weather than on the skill of the angler. The shallowness of the sounds makes them particularly susceptible to winter cooling. Nowhere is winter's effect more evident than in the broad Pamlico Sound, which because of its long fetch—the distance over which the wind is blowing—can experience a marked drop in temperature during a period of persistent, cold northwest winds. One sound species that is especially affected by cold weather is the spotted seatrout. Research has demonstrated that seatrout start to become lethargic when the water temperature drops into the upper 40s. When the mercury dips to 45°F or lower, mortality is a distinct possibility for this species. Records show that a big winter kill occurred in January 2003, setting the trout population back and requiring three years for it to recover. If the onset of the chill is rapid, seatrout may not have time to react; they can become trapped in the shallow creeks that border the sound and might be unable to seek the shelter of deeper sound water.

In contrast, summers along the Outer Banks can be brutal, with sound temperatures spiking well above 75°F. Fish can become extremely lethargic, consistently not responding to any outside stimuli. As members of a cool-water species near the southern end of its range, resident young striped bass are especially vulnerable to high temperatures. But if fish are caught during periods of high water temperature, they may suffer increased physiological stress, greater oxygen debt, and a spiraling upward mortality rate. Also, the amount of dissolved oxygen in the water decreases with increasing water temperature. (Water at 85°F will have only half the dissolved oxygen of that at 32°F.) My recommendation is simple: when water temperatures peak, both the duration of the fight and the handling time should be minimized. The length of time a fish is "played" significantly increases physiological stress on the fish due to depletion of its energy resources and accumulation of lactic acid within its muscle tissues, which can result in extreme fatigue. The buildup of this noxious substance is the result of the loss of oxygen in muscle tissue while under physical exertion. Due to the lower dissolved oxygen content in warm water, fish caught in this water will build up lactic acid faster than fish caught in cold, oxygen-enriched water. Your quick release will increase the chances of the fish's survival, allowing it to fight another day.

TARGETING FISH: TACKLE AND TECHNIQUES

Leave the twelve-foot surf rod back at the house; long casts and heavy weights are not needed to catch many of sounds' denizens: spotted seatrout, red drum, flounder, bluefish, and weakfish. Except for fighting 100-plus-pound tarpon and big red drum, a light casting or spinning outfit (eight- to twelve-pound class) will suffice for most situations.

When targeting seatrout, most anglers prefer a leadhead jig (one-eighth to one-fourth ounces) adorned with a soft plastic, curly-tailed grub. Heavier jigs (three-eighths to one-half ounces) are appropriate when the current is running strong or you are fishing the deeper sloughs. In particularly shallow water, anglers suspend the grub a couple of feet below a float. Reeling in the float after the cast creates a surface commotion—a popping, gurgling noise—which is the marine equivalent of ringing the dinner bell. Relatively recently, many guides have switched to artificial scented baits, like Berkley Gulp! (shrimp being a favorite), on their jigs. The baits are made of biodegradable ingredients and look, feel, and smell like the natural species. While these baits are effective on trout, even pinfish, pigfish, and crabs enjoy making a meal out of them.

The sounds are also ideally suited for fly-fishing because of their relatively shallow waters and, best of all, lack of crowds. Along the Outer Banks, fly-fishing attracts fewer participants than surf or pier fishing, and the local tackle shops do not stock rods, reels, or even a selection of flies. Be prepared to bring all your fly-fishing tackle.

A seven- to nine-weight rod will allow you to cast even bulky flies when the wind is blowing. While trout fishing on freestone streams or spring

Fly-fishing flies (left to right): deceiver, Clouser minnow, and shrimp (author's photo)

creeks requires floating lines, sound fishing calls for a fly line that gets the fly down to where the fish are feeding. Manufacturers classify sinking lines depending upon their rate of descent. An intermediate fly line, which sinks at a rate of one to two inches per second, allows the angler to retrieve a fly in a very precise or narrow depth range. For example, if the sea grasses are three to four feet deep, an intermediate line will keep the fly just above the tops of the vegetation. As with any sinking line, the longer you wait, the deeper the line will sink, so you should be prepared to start your retrieve when the fly has reached the strike zone. Fast sinking lines (three to five inches per second) are ideal for fishing drop-offs or channels, where fish may be hunkering down near the bottom. An eight- to ten-foot leader, tapering to a ten- to twelve-pound tippet, will handle small red drum, trout, and bluefish. At the business end of the tippet, the flies of choice are Clouser minnows, deceivers, and shrimp patterns. I am particularly fond of Clouser minnows; though they were developed for use in fresh water, dozens of saltwater species have succumbed to the lure of these flies. In 1984, Bob Clouser attached metallic eyes to a deer-hair-dressed hook to produce a weighted fly that mimics the action of a baitfish constantly darting to escape a predator. In addition, attaching the eyes to the hook turned the fly upside down in the water—preventing the fly from snagging on debris.

With moving schools of fish, anglers must be prepared to deliver a fly quickly and get it down fast. Ideally, the fly should be in front of the school at approximately eye level. Fish that are moving fast through the target zone will rarely be looking up, so sinking lines and big flies are needed.

While a fly reel should have adequate drag for those times when a beefy fish makes a determined run, it is not used to retrieve the fly. After the fly lands in the target area, most anglers grab the fly line just above the reel and employ a stripping technique (short pulls on the line) to impart action and movement to the fly.

LOCATING FISH:
CHANNELS, GRASS BEDS, AND WEED FLATS

Since these vast estuaries hold more than 2,500 miles of shoreline and 3,000 square miles of water, locating fish in them can be quite daunting to many anglers. Hiring an inshore guide (many are based out of the marinas) is the quickest way to learn the area. If you prefer to strike out on your own, I recommend obtaining a navigational map of the area from a commercial dealer (Maptech) or from the National Oceanographic and Atmospheric Administration. These maps depict the sounds' depths, underwater hazards, and to some extent, even shoreline aquatic vegetation. I have recently started to use the computer program Google Earth to study these sounds. This application can zoom in on many specific sites, providing detailed images of shoreline features and depths. Movement of the computer's cursor across the image matches changing latitude and longitude coordinates. Find a promising spot, and you can save these coordinates on your Global Positioning System.

Successful sound fishing entails searching for prime fish habitat: grassy weed flats, channels, submerged grass beds, stump fields, and marsh sloughs. The rich bottom grasses that blanket parts of the sounds attract shrimp, anchovies, and crabs that during the summer months are preyed upon by trout, bluefish, and flounder. Deep channels and sloughs provide conduits for fish to navigate the grass flats and also ambush prey. In contrast, many of the sandy shoals or big sand flats are generally devoid of game fish because their lack of cover means little food. To be a successful sound angler, keep in mind a simple, but true, maxim that I heard long ago from a biology professor: critters go where there is food. Look for the sites where prey will congregate.

Wading the shallows of the sounds is not quite the equivalent of stalking the sunlit flats of the Bahamas. The sounds' murky waters are not conducive to sight casting to a cruising fish. Repetitive, blind casting—covering as much water as possible—yields the best results, assuming the presence of a fish-holding structure, such as a slough or bottom depression. The structure need not be big. I've had success fishing small, inconspicuous holes no more than fifty to sixty feet in diameter.

I often target flounder, which settle on the bottom, waiting for their next meal to sweep past them. My technique is relatively simple: cast to the edge of the hole, let the fly sink to the strike zone, and then provide a little erratic motion to the fly that, hopefully, imitates the movement of a wounded baitfish. Upon the hookup, I always keep a tight line to the fish. Flounder are notorious for swimming toward you, giving the impression that they have thrown the hook. While my attempt to find these isolated structures has often entailed wading a quarter-mile or more from the shore, the reward of one of the sound's treasures at the end of my line has been well worth the effort.

While the warm waters of summer preclude the use of waders, be aware of stingrays lurking on the bottom. Shuffling your feet as you wade out from the shore is strongly advised to send these painful nuisances scurrying away.

If you are not inclined to this seek-and-search routine, there are a number of locations that are well known to both wader and boater alike.

Manns Harbor Bridge

Spanning the Croatan/Albemarle Sound, the bridge connects the northern part of Roanoke Island to the mainland, and over the years it has become a focal point for anglers targeting striped bass. This fishery is approached mainly from shallow-draft boats because most of the shoreline is not easily accessible.

Except during the dead of winter, anglers can expect school-size stripers almost year-round, with the best action generally between late October and December. With cooling temperatures in the fall, stripers often feed actively on the surface and can be caught by casting top-water plugs on light spinning outfits or fly-casting deceivers, poppers, or sliders around the bridge structure. Tidal flow is generally weak in this area, but fish seem to be more active in the presence of wind-driven currents.

Melvin Daniels Bridge

This bridge on the Manteo–Nags Head Causeway (connects Nags Head to Roanoke Island) allows anglers to pursue a wide variety of Roanoke Sound species, including spotted seatrout, croaker, spot, and weakfish. The area has ample parking for approximately thirty vehicles, a well-lighted pier, and a fish-cleaning table. Most of the surrounding area is quite shallow, less than one to two feet in depth, but near the western end of the bridge, there is a navigation channel (three feet deep) that is marked by white poles. This channel continues all the way to Oregon Inlet and is clearly visible on the images provided by Google Earth.

Off Island Slough

Near the Bodie Island Lighthouse, a few miles from the north entrance to the Cape Hatteras National Seashore, is a channel that stretches several hundred yards in a north-south direction. Anglers access the channel by parking in the visitor center lot and walking down a gravel road at the south end of the parking lot to a small private dock. The channel separates the dock from Off Island. Though there are areas with a soft, muddy bottom along the channel, wading is generally not a problem. Currents in the center of the channel often run strong, so an intermediate fly line or a light (one-quarter ounce) jig may not be sufficient to reach the fish holding in the deepest areas of the channel. Presenting your offering close to the bottom is the key to consistently catching red drum and speckled trout during the day.

Green Island Slough

Within the fraternity of wading anglers, this channel ranks high on their list as one of the most productive fishing locations in the seashore, holding almost every type of sound game fish. The slough is located at the south end of the Bonner Bridge, and anglers can park in the adjacent lot and walk across NC 12 to the sound. Because of the slough's proximity to Oregon Inlet, tidal currents run strong, and the bottom can be quite unstable near the bridge. Green Island Slough is long, stretching more than 1,000 yards to the southwest, and deep, ten feet or more in places. Because of the sharp drop near the edge of the channel, anglers avoid wading too close to the slough and opt for the longer cast. A forty- to fifty-foot cast from waist-deep water with a fast sinking line will land your fly in the strike zone. If you can't make this cast, opportunities exist during clear-water conditions to wade closer to the channel, since the edge of the deep water is easier to see. Uncomfortable with fly-fishing tackle? Switch to a seven-foot spinning outfit and cast a MirrOlure plug to a trout, let it sink, and slowly retrieve the lure with a couple of stops. Many anglers only target the channel during low tide because the ebbing water in the sound often concentrates the fish in the slough.

New Inlet

Approximately seven miles south of Oregon Inlet is a boat-launch site perfect for kayaks and johnboats. Though New Inlet was indeed an inlet, first opened around 1738, it has been closed since 1933. When Oregon Inlet opened in 1846, the land between New Inlet and Oregon Inlet became known as Pea Island. But the opening of the wider Oregon Inlet meant less water flow through New Inlet into Pamlico Sound, resulting in its closure. What remains is an old channel that can be accessed by wading north from the boat-launch area to the base of an old wooden bridge that once spanned the inlet. The slow current here makes this channel somewhat easier to fish than Green Island Slough.

Canadian Hole

Just south of Avon, Canadian Hole is an area where Hatteras Island was breached by storms in the 1960s and 1970s. Now a popular windsurfing area, Canadian Hole attracts its fair share of anglers because of its deep water and minimal tidal flow. Plan your fishing activities during the early morning or after sundown when the windsurfers have left the area to the anglers. Depending on the time of the year, weakfish, speckled trout, croakers, and small red drum are all caught here by both fly and spin anglers.

Western Pamlico Sound

The western side of Pamlico Sound is composed of marsh islands, shoals, and oyster beds, all harboring an abundance of life. During the summer, red drum prowl the area, consuming crabs, menhaden, and mullet. Under calm conditions, anglers have numerous opportunities to sight cast to these fish, employing either spinning or fly-fishing outfits. Summer is also prime time for tarpon, but speckled trout can be found within the sea-grass beds.

Hatteras, Ocracoke, and Drum Inlets

Two target sites are the sand shoals (flood tidal deltas), located on the sound side of these inlets, and the sloughs within the sound. The former are formed by the flooding or rising tide carrying water and sediment through the inlets and depositing the sediment load in the sound. Flood tidal deltas are constantly being reworked by the incoming flow that tends to follow different paths across the delta, resulting in a complex pattern of shoals and small channels. While these sand shoals are generally not productive, red drum will occasionally move up onto these sites. In this relatively shallow water, anglers can get some sight-casting opportunities to these reds, but they are, not surprisingly, very jittery and will move into deeper water when they feel threatened. But during low-light conditions, like early morning and evening, the red drum may not readily spook and might allow anglers a shot or two at them.

The shoals of Ocracoke Inlet are particularly known for their drum fishery in the spring. Here anglers throw three- to five-ounce bucktails on spinning rods or cast four- to five- inch-long flies that sink quickly and have the action of a jig. When the water clarity decreases, anglers switch to bait fishing.

In contrast to the shoals, the sloughs will hold and concentrate trout. Look for drop-offs four to eight feet deep, where the eelgrass and cordgrass flats tumble into deeper cuts or channels.

North Core Banks

Probably not as well known as the above locations, this site is approximately two miles north of Old Drum Inlet. Both north and south of the ferry channel are two relatively deep depressions that can be accessed using the National Park Service beach cabins as a staging point. It is a short paddle by kayak to reach these holes, and both can at times hold trout and flounder. Since the north hole is smaller than its counterpart to the south, it requires a more stealthy approach. If these sites are not holding fish, paddle northward. There are a number of sloughs and holes that are quite close to the shoreline.

FISHING THE SEASONS

The sounds offer anglers a year-round fishery. During the spring, when the shallow grass flats warm up due to the longer days, spotted seatrout (specks) and weakfish begin to make a showing. The trout, particularly the specks, will move into the flats but may be spread out over these large mats of vegetation. As spring transitions into summer, red drum move into the sounds through the inlets to breed, and anglers may encounter fish of over forty pounds. Late afternoon and night fishing are best for catching these brutes. Tarpon fishing, particularly in Pamlico Sound, heats up in the middle of the summer, and bluefish and flounder also abound. High summertime water temperatures will send the blues scurrying to the deeper, cooler waters of the sounds. During the fall, small (five to ten pounds) red drum are feeding heavily prior to their migration to the ocean. These fish move in relatively large schools, and anglers catch their fair share almost every day. In addition to the reds, spotted seatrout and weakfish can also be taken. Throughout October the specks are schooling, preparing to make their run to the ocean. They are moving en masse across the sound, heading to the inlets. By winter, the signature species is striped bass, and anglers turn their attention to hungry stripers that are willing to hit a big fly, bucktail, or plug.

THREATS TO SOUND FISHING

A sustainable recreational fishery in North Carolina depends on the health and vitality of its estuaries, which are part of a larger system: a watershed that includes the rivers and creeks that flow into them and the land surrounding these streams. The drainage from the mainland watershed has had, and continues to have, a negative impact on water quality. One particularly insidious problem is cultural eutrophication—physical and biological changes that occur when excessive nutrients enter the aquatic environment from non-point sources, such as fertilized lawns, fields, and farms. These nutrients, mainly nitrates and phosphates, stimulate the growth of algae that can result in massive blooms of these microscopic plants. The algae cloud the water, block sunlight to submerged sea grasses, and ultimately cause the grasses to die.

In addition, when the algae die and decompose, they use up dissolved oxygen that is needed by countless organisms in the sound. Large areas of eutrophication can result in extensive hypoxic (hypo = under, oxid = oxygen) dead zones that often occur at the mouths of rivers after large spring runoffs. Oxygen levels within these zones drop from more than five to less that two parts per million—an amount that is lower than most organisms can tolerate. While the more mobile organisms are able to flee the area, many sedentary bottom organisms succumb to the hypoxic condition.

Offshore Fishing

The sport of offshore fishing along the Outer Banks started on Hatteras Island in 1937, when Ernal Foster, a longtime resident of Hatteras, conceived of the idea of building a boat that could be used for both commercial and recreational fishing. With it he could fish the winter months with fellow watermen and pursue big game fish the rest of the year. His original boat, the *Albatross*, which cost only $805, was to become one of three boats built between 1937 and 1953 that comprise the Albatross Fleet. The durability of these boats and the commitment of those who captain them stand as a testament to the fact that this offshore fleet has fished every season except during the years of World War II. But initially the charter business was slow to catch on. In 1937 Foster made only four trips, at a price of $25 per charter. Though the Albatross Fleet still fishes on a regular basis, offshore fishing in the twenty-first century has changed considerably since Foster's time.

The original *Albatross* was made of wood, mostly juniper, which Foster gathered and cured himself to fashion his boat. The vessels of today sport fiberglass hulls, state-of-the-art electronics, and price tags that can reach seven figures. While in 1937 Ernal Foster was the only Hatteras islander to charter his boat, there are, at present, dozens of boats to choose from.

The price of an offshore trip may run as high as $1,400, but most marinas offer the option of a makeup charter, a maximum of six anglers willing to share the expenses. These include $220 to $250 for the charter; a tip—usually 10 to 15 percent—to the mate; and the cost of cleaning any fish that are caught during the trip. Turns at the rod are rotated among all the anglers, and each angler shares in the total catch.

Boats leave the marina at approximately 6:00 A.M. and return to the dock by about 4:00 P.M. The run to the fishing grounds may be slightly over an hour to almost two hours—still leaving plenty of time for fishing. To many anglers, offshore fishing in North Carolina generally means a jaunt to the Gulf Stream, which on the average is about twenty miles from Cape Hatteras and about twenty-five miles from Hatteras Inlet.

The Gulf Stream is a powerful, warm, and fast Atlantic current that originates in the tropics, flows northward along the eastern coastline of

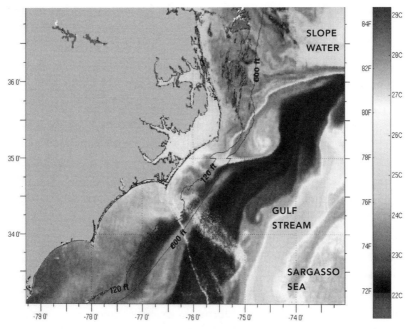

Infrared image of Gulf Stream (courtesy of NOAA)

the United States, and is part of the large, clockwise-flowing North Atlantic gyre. In 1855 Matthew Fontaine Maury, the "father of modern oceanography," wrote of the Gulf Stream, "There is a river in the ocean." Standing on the shores of the Atlantic, an observer can detect no sign of this river of water. But just over the horizon flows this vibrant and dynamic force of nature within well-defined boundaries, characterized by dramatic changes in physical and chemical properties. Within these watery boundaries, like the banks of a continental river, the blue, clear waters of the Gulf Stream stand in sharp contrast to the dull, grayish-green coastal waters. In addition, sea surface temperatures in the Gulf Stream are relatively warm—above 80°F and at least fifteen degrees higher than outside the current. As we will soon see, temperature, more precisely temperature contrast, is a key variable in locating fish.

TARGETING FISH: TACKLE AND TECHNIQUES

The majority of the game fish that anglers seek in the Gulf Stream are large, pelagic predators, such as tuna and marlin, which cruise the current in search of prey. While trout may have particular feeding stations and holding sites in a river, the nomadic game fish of the Gulf Stream rarely—if ever—permanently associate with specific bottom features or structures.

The most effective method for covering large areas of the open ocean in the least amount of time is trolling. In its simplest form, trolling utilizes a boat to pull bait or a lure—or a combination of baits or lures—through the water to attract and catch fish. Charter boat captains may troll six or more lines, which constitute a trolling spread, that can vary in configuration and composition depending on the game fish being targeted. Like a spider web, these lines radiate to the port, starboard, and stern of the boat. To obtain maximum spread of the lines and avoid their tangling, outriggers are used. Outriggers are a pair of aluminum or fiberglass poles fitted on each side of the boat at approximately a seventy- to eighty-degree angle. The outriggers have release clips attached to a nylon line running along the pole like a halyard on a flagstaff. The main line from the reel is attached to the release clip that, when under pressure from the strike of a fish, snaps open to allow the angler to fight the fish. From the boat's stern, most anglers employ at least one flat line into the water, which because of its low angle minimizes tangling and keeps the bait or lure just below the surface.

The keys to trolling success are placing the bait or lure where it is most visible to the game fish and having it move through the water in a manner that is attractive to the fish. Most fish have particular feeding habits and mannerisms that will dictate whether they will strike a lure or bait, or spurn the offering. The game fish is biologically programmed to view the bait in a specific way. If the appearance or movement of the bait appears unnatural to the fish, the angler's chance of catching the fish is minimal. The angler, or for sure the captain, must be knowledgeable of the nature (type, size, color) of the bait or lure that the fish is seeking and present it in a natural manner. It must be positioned at the proper depth and its movement controlled through the water.

Trolling can be divided into a combination of four simple categories: slow, fast, shallow, and deep. Though the Gulf Stream waters are several hundred feet deep, shallow trolling positions the bait at or near the surface. This method is particularly effective on billfish or dolphinfish (mahi mahi) that feed on schools of bait located near the surface. Flying fish and other small baitfish swim close to the surface, and they are relentlessly pursued by Gulf Stream predators.

Even novice anglers can recognize a flying fish by its winglike pectoral fins and lopsided tail—the lower lobe is larger than the upper lobe. The escape mechanism for a flying fish is to swim rapidly to the surface and propel itself by means of its powerful tail into the air to avoid the massive maw of its pursuer. Once airborne, flying fish may glide more than 300 feet across the sea surface.

Since the trolled bait must closely match the natural bait, dead ballyhoo are routinely used on most Gulf Stream charters. (Ballyhoo are easier to rig than flying fish, relatively cheap, and readily available at most bait and tackle shops.) Since most ballyhoo come frozen, the mate must first thaw them before they can be rigged for trolling. Remember, to elicit a strike from a hungry predator, the ballyhoo must appear natural when it is in the water. To achieve that end, the mate often literally squeezes the feces out of the ballyhoo, making it more pliable, before inserting the hook through its cavity. Ballyhoo may be rigged either naked or dressed, using nose cones or skirts on the bait. Many anglers believe that a colorful skirt attracts fish into the trolling spread, and the cone may create an appealing bubbly trail behind the ballyhoo.

A fast troll—six to eight knots—"skips" the shallow running bait across the sea surface and is the main method employed by most North Carolina captains. But flexibility and adaptation by the captain to what the Gulf Stream offers up on any given day may be the difference between boating fish or telling fish stories back at the dock. For example, if the current is flowing against a brisk wind, resulting in short and choppy seas, slower trolling speeds are in order to keep the bait in the water. On the other hand, fish will sometimes show minimal interest in the bait, following it but refusing to strike. To trigger their feeding instinct, the trolling speed is increased to make the bait swim faster and skip more energetically. Often this technique may draw a strike because, at its core, it simulates a prey that is trying to escape.

If anglers are targeting billfish, they may add teasers to the trolling spread. Teasers—artificial lures with no hooks—attract fish to the boat and excite them to strike the bait. (They have also proven effective on non-billfish species.) The splashing and/or flashing action of the teasers is thought to act as a strong stimulant to a wary fish. Teasers fall into two categories: bird teasers and daisy chains. The former are wooden or plastic lures that have a wing on each side of the lure's body. The wings give "life" to the body. The result is a lure that mimics a flying fish that is attempting to flee a predator. Bird teasers can be trolled separately or accompanied by a single hooked

Charter fleet out of Hatteras (author's photo)

lure or string of lures behind them. Trolling at four to eight knots in calm to moderate seas appears to give the best action to these teasers. Daisy chains, which are generally a combination of plastic squid, soft-headed lures, and bird teasers, are trolled from either side of the boat, with the bait or lure set on the outside of each chain.

Trolling shallow and slow generally means employing live bait, such as menhaden, cigar minnows, and striped mullet. Since success depends heavily on the ability of the live bait to swim, at least to some degree, the skipper trolls as slow as the engine will allow, keeping the bait barely moving behind the boat.

Deep trolling affords the angler the opportunity to target fish that may be holding well under the water surface. It can be achieved in several different ways. The North Carolina charter fleet commonly employs planers to get the bait down to the desired depth. As the angler tows the planer behind the boat, the flow of water over the planer forces it to dive. (The larger the planer, the deeper it will descend, due to its large surface area.) When a fish strikes the trailing lure or bait, the ring that the fishing line is attached to slides forward and "trips" the planer. At this stage, the planer is in line with the fishing line and the lure. This configuration alleviates any water pressure on the planer and allows the angler to fight the fish. While most trolling situations call for fifty- to eighty-pound class tackle and the appropriately sized monofilament line, the use of planers generally necessitates using wire line to achieve the desired depth, but because of its small diameter, braided line is also being used more by offshore anglers.

Once the mate puts out the trolling lines and the captain sets a course, it is time for most anglers to sit back and relax. Trolling involves long periods of inactivity interspersed with moments of frantic action when a fish appears in the trolling spread. Under the watchful eyes of the captain and the mate, a flash of color streaking across the water is a sign to them that a game fish has taken an interest in their offering. The "popping" of the release clip frees the line from the outrigger pole and signals the start of fishing's version of a Chinese fire drill. The captain barks out orders to clear all lines (reel in the other trolled baits to avoid tangling with the hooked fish), man the fighting chair, and, at all times, keep pressure on the fish. If all aboard can choreograph these tasks smoothly, then their reward is a hard-fighting fish that ultimately yields to the gaff of the mate.

One specialized method of trolling—though some may argue that it is not technically trolling in the true sense—is kite fishing, which can be particularly effective on tuna. At the minimum, two rods are needed—one from which the kite is flown and the actual fishing rod. A release clip slides up the kite line and holds the fishing rod's line that has a lure or bait hanging down into the water. When a fish strikes, the fishing line is pulled from the kite, allowing the angler to fight the fish. With a kite, the leaders are out of the water, and the kite can be positioned at almost any distance behind the moving boat. The key to successful kite fishing is positioning the bait or lure so that it skips across the surface; it should touch, splash, and hop, but not dive below the surface. Since kite fishing requires a special rod from which the kite is flown, careful maneuvering of the boat, and adequate wind, it has not been widely employed by novice anglers.

Even more intricate than kite fishing is the use of green sticks in the tuna fishery. Commercial Japanese fishermen introduced green sticks to Hawaii in the early 1980s, but they have also proved successful off the east coast of the United States. The term "green stick," which originated from the composition of fiberglass resin and material that created a green color, now applies to all heavy-duty poles (thirty-five to forty-five feet tall) regardless of color. The green stick, which is mounted vertically just aft of the cabin, elevates the main line and tows a twenty- to thirty-pound wooden bird about 300 yards behind the boat. The bird keeps the main line taut and acts as a teaser to attract fish to the surface. Dangling from the main line between the boat and the bird are four to six drop lines with attached plastic squid lures. If positioned correctly, the squids will skip along the surface, dipping in and out of the water. This motion often proves irresistible to a hard-charging tuna, which has been known to launch itself completely out of the water in an effort to devour the squid. Each of the drop lines is linked to the main line with a breakaway feature, such as a release clip. When a fish strikes, the drop line breaks from the main line, so that a fish can be fought with a rod and reel. Since green stick fishing requires heavy tackle, it may not be the most sporting technique to catch tuna, but it definitely is effective in putting fish in the boat.

Though trolling is at the essence of Gulf Stream fishing, to the untrained eye, it may appear that these lures and bait are skipping over a seemingly endless expanse of homogenous water. But the key to success is spatial heterogeneity—observable or measurable changes in particular physical or biological factors. Simply put, the "place" is at the essence of finding fish.

LOCATING FISH: EDDIES, FINGERS, AND FRONTS

The location of game fish is determined by two factors: the comfort of a particular species and the availability of food. Water temperature affects the well-being of a game fish and the concentration of prey. Each Gulf Stream species has a water temperature range it prefers and a wider range it can tolerate without stress.

Though the Gulf Stream is more than sixty miles wide, savvy captains target particular sea surface "structures," which meet the needs of the species, to find fish.

Thermal Fronts

As the Gulf Stream approaches the North Carolina coast, the temperature gradient between the coastal water and the western boundary of the Gulf Stream is more pronounced than along any other portion of its path. This sharp change in temperature over a relatively short distance is a thermal front, analogous to a weather front in the atmosphere. The sea surface temperature near a thermal front may change more than twenty degrees in only a few miles. To a Gulf Stream game fish, thermal fronts mean food, plenty of it. According to Dr. Mitch Roffer of Roffer's Ocean Fishing Forecasting Service, "If this food chain exists within a game fish's preferred temperature, it's only a matter of time before they find it and stack up in the area." Thermal fronts, especially those with well-defined edges, can act as barriers to the movement of game fish and their prey. Seeking their thermal comfort zone on the warm side, these fish will move along the edges instead of crossing them. In addition, because large concentrations of plankton cloud the water, sight-dependent hunters, such as marlin and tuna, will tend to stay on the warm side of the front.

The location of thermal fronts may vary over time and with distance from the coast. To help locate thermal fronts, some captains may employ sea surface temperature charts that have been processed from satellite data. Specifically, satellite sensors measure the amount of heat or, more accurately, the infrared radiation given off by the ocean surface. Since the amount of radiation emitted by an object is proportional to its temperature, researchers can accurately determine the sea surface temperature by measuring the amount of radiation emitted from the ocean. The resultant product (chart) is a color-coded image that depicts the water temperature in relation to the geography of the area and depth of the water.

Rings

Though the Gulf Stream flows in a general northward direction along the eastern seaboard, this current often traverses the North Atlantic in a snake-like path. As if alive, the Gulf Stream twists and turns in great sweeping loops or meanders. At times, a meander bends sharply and forms almost a complete loop. This pinched loop may separate from the main flow, much as an oxbow lake forms from a meander in a river, resulting in a closed, self-contained vortex known as a ring. As each ring develops, a column of water is appropriated from one side of the current and transported to the other side—a region of distinctly different physical, chemical, and biological properties. Warm-core rings, for example, are large (100 to 200 miles in diameter), clockwise-circulating masses of water that are found inshore of the Gulf Stream and are surrounded by colder water. In contrast, cold-core rings are pockets of counterclockwise-flowing cold water surrounded by the warmer water of the Sargasso Sea. Because warm-core rings are home to a variety of tropical and subtropical planktonic species, a food chain develops in a relatively short time. Small baitfish, which are attracted to the cornucopia of plankton, become the prey of cruising game fish.

Though most large warm-core rings initially form north of Cape Hatteras, many of these rings over their life span of six months move west to southwest. As they travel, they bounce between the shallow continental shelf and the Gulf Stream, decrease in size, and ultimately are absorbed by the Gulf Stream. Due to their proximity to the coast, these warm-core rings are frequent targets for offshore anglers, who may be reluctant to make the longer trip across the Gulf Stream to the cold-core rings.

Fingers

The difference in current speed along the Gulf Stream's western boundary often creates elongated filaments of warm water, or "fingers." Since these fingers are connected to the Gulf Stream at their northern end, the current continuously pumps warm water into the fingers. During the spring and fall, when sea surface temperatures are relatively low west of the Gulf Stream, game fish and bait that are migrating along the western edge of the Gulf Stream are often found concentrated in these fingers, but in turn, they avoid the relatively cooler water that gets entrained between the fingers and the Gulf Stream.

Frontal Eddies

Along its western boundary, the Gulf Stream often sheds rotating masses of water known as frontal eddies, which are considerably smaller in size and have a shorter lifespan than rings. These eddies are indicated by changes in water temperature, current direction and speed, and water color.

A counterclockwise (cyclonic) rotating eddy forms when a southward-projecting finger entraps a mass of cold water between itself and the Gulf Stream. The rotation of water around the eddy results in an outward flow of surface water, which is ultimately replaced by subsurface water, in a process known as upwelling. This upwelling eddy infuses the sunlight layer with the nutrient-enriched bottom water and ultimately results in a concentration of prey feeding on the nutrient-dependent plants. While most anglers prefer trolling in the same direction as an eddy's current since the bait will travel with the flow, others will probe the center of the eddy to assess its productivity.

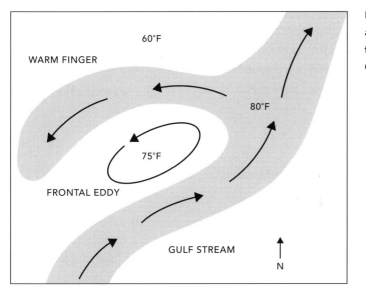

Frontal eddy and warm finger of the Gulf Stream

A clockwise (anticyclonic) rotating eddy forms when a portion of a finger separates from the Gulf Stream. While this pocket of warm water lacks both high-nutrient and phytoplankton concentrations, during the spring, pelagic species may concentrate here, since the surrounding water remains cold.

Rips

These features are long, narrow bands of calm water and are bracketed on either side by rough water. Floating weeds and debris will often collect in the calm areas of the rip. In particular, sargassum weed (actually a brown alga) frequently aggregates into long, thin rows, or flotsam lines. Gulf Stream sargassum supports a diverse assemblage of organisms, including fungi, micro- and macro-epiphytes (plants that grow on other plants), and dozens upon dozens of invertebrates and small fish. This web of life is a "floating jungle" that attracts apex predators. With its tangled maze of stems, fronds, and floats, sargassum also provides shade (at least, that is the prevalent thinking by many anglers) for some game fish, especially dolphin, and refuge from bigger predators, such as blue marlin.

Before beginning to fish, most knowledgeable anglers will survey a weed line to see if bait is holding within the sargassum. Weed lines that are barren seldom prove productive in attracting game fish. If a weed line appears promising, anglers troll lures along the edges of the rip. Because the Gulf Stream flows north, and many game fish swim with the current as they migrate in search of fertile feeding grounds, trolling with the current give the lures a more natural appearance, as they become "eye candy" to a cruising game fish.

THE SARGASSUM COMMUNITY

Because of the unique assemblage of organisms within the sargassum community, which attracts predatory game fish, the South Atlantic Fishery Management Council (SAFMC), the federal agency responsible for protecting pelagic fish and their habitat from North Carolina to Florida, has declared sargassum as "essential fish habitat" and is charged by law with minimizing any adverse effects on such habitat. But who are the challengers to this ecosystem? Dating back to the 1970s, North Carolina–based fleets had routinely harvested sargassum, albeit in small amounts, for use in nutritive supplements and feed additives. Concerned that the harvesting would escalate to a much greater level, which would negatively affect the abundance of sargassum, the SAFMC developed and implemented in 2003 the Sargassum Management Plan, which places strict limits on the harvesting of sargassum within 100 miles of the coast. Though this was a bold conservation step, the ability to secure protection for sargassum in international waters is less sure. The Sargasso Sea, where most of the sargassum that enters the Gulf Stream originates from, is vulnerable to exploitation at unsustainable levels because it resides in the essentially lawless realm of the high seas. On the positive side, the distance needed to travel to reach the bulk of the sargassum weed may make any large-scale harvesting venture economically unappealing.

FOLLOW THE BIRDS

There are times—we've all had them—when the fishing goes south; all types of bait, lures, and teasers are routinely spurned by the denizens of the Gulf Stream. What is a frustrated captain to do? One can't go wrong in imitating the Admiral, Christopher Columbus, himself: follow the birds. On

October 7, 1492, Columbus had been at sea for almost a month and still had not come upon the rich lands that he had promised his crew. Mutiny was brewing; the crew no longer had the stomach for Columbus's quest. But by evening Columbus observed a large flock of birds flying from the north to the southwest. He interpreted their flight pattern to mean the birds were either heading to land to nest for the night or migrating to warmer lands in response to the approaching winter. He soon altered his course from west to west-southwest. Determined to follow his instincts, Columbus doggedly held this course, and his persistence paid off; he sighted land four days later.

In close proximity to the North Carolina coast, the interplay of the warm waters of the Gulf Stream with the cold water on its western edge supports a diverse group of seabirds, including shearwaters, terns, and the black-capped petrel—one of the signature species of the Gulf Stream—which feed on baitfish that attract dolphin, wahoo, and tuna. In particular, yellow-fin tuna, using the current as a migration route, routinely bust baitfish that they have driven to the surface. The water "boils" with activity: fish slash through the bait, and screeching birds swoop down to grab their share. Skippers and mates scan the horizon in hopes of detecting flocks of birds that are working above these feeding frenzies. Upon marking a flock and ascertaining its direction of travel, the captain will motor ahead of the flock to intercept its path and, hopefully, a school of hungry fish. When the boat is in position, the mate deploys the trolling lures, and the action comes quickly. Trolling through this triangle of birds, bait, and tuna often elicits multiple strikes and hookups from the latter. Most of the tuna caught will be in the same size range. Some schools are composed of thirty-pounders; others have tuna approaching sixty or more pounds.

THREATS TO GULF STREAM FISHING

In South Florida, researchers from Florida Atlantic University (FAU) are already working on a project to harness power from the Gulf Stream. Their idea is to employ a series of underwater turbines moored in the heart of the current, where it has a flow rate more than 300 times that of the Mississippi River.

Though remote, negative environmental consequences must be factored into the project because though the prototype turbine rotors are only 10 feet in diameter, the full-sized turbines will have rotors 100 feet in diameter. What effect would a grid of these large turbines have on migrating fish? What level of mortality, if any, is acceptable?

The researchers are also looking to see what effect the turbines could have on the flow and strength of the Gulf Stream. This current is part of something much larger than itself—a global "ocean conveyor belt." Though the details of this conveyor belt still need to be worked out, the Gulf Stream represents the upper, or surface, limb of the conveyor, which also includes deep flows that transport water throughout the world's oceans. A decrease in northward transport by the Gulf Stream could potentially affect the whole ocean conveyor belt, disrupt latitudinal heat transport, and trigger a climatic shift, which could then displace various species. The stuff of science fiction? Currently, FAU's researchers believe that the turbines will have little impact on the overall strength of the Gulf Stream. But the most negative scenario shows that at peak generation, approximately one-third of the current's total energy will be extracted.

Inshore Fishing

Inlets are the physical connections between the sounds and the inshore waters. Exactly where one of these environments ends and the other begins is open to debate, so in this chapter I will also view parts of the inlets as falling within the inshore realm. In particular, on the ocean side of inlets are ebb tidal deltas, the counterparts of the sounds' flood tidal deltas. These large sandy deposits result from the outgoing tide carrying sediments through the inlet. Over time, these shoals eventually drift southward with the coastal current and become part of the northern portion of a barrier island. The large sand flat on the northern end of Portsmouth Island is a prime example of the dynamic change occurring at the mouth of Ocracoke Inlet.

Seaward of the inlets is the continental shelf—the extended perimeter of the continent—which was part of the land during the last glacial period but is presently undersea. The shelf has a gently sloping bottom that reaches from the coastline to the shelf break, a marked increase in offshore depth. For example, the shelf off the Outer Banks extends seaward in a gentle incline, about nine feet per mile, or much more gradual than the slope of a well-drained parking lot. Near Cape Hatteras, the shelf stretches approximately eighteen miles out into the Atlantic and ends at about a depth of 250 feet.

The continental shelf is covered by terrigenous sediments, that is, sediments derived from erosion of the continents. Little of the sediment, however, is from current sources; the bulk of these deposits, known as relict sediments, occurred during the last ice age, when sea level was 300 to 360 feet lower than it is now. At present, wind-driven currents most assuredly reshape the bottom sediments, but in general, the shelf lacks pronounced changes in underwater topography. New bottom structure can form overnight when storm-generated waves roll in over the shallow shelf. Based on wind conditions (speed, duration, and fetch, or the distance over which the wind blows), wave forecasters can estimate the resultant wave properties, such as height, length, and period. (For example, with a wind speed of 35 miles per hour, a duration of 23 hours, and a fetch of 300 miles, the waves, on the average, would have a height of 14 feet, a length of 250 feet, and a period

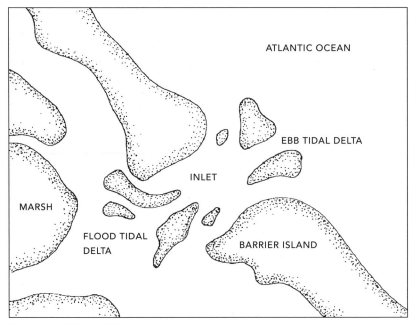

Barrier island inlet and tidal deltas

of 9 seconds.) As waves move across the ocean surface, their presence is also felt below the surface, and the key variable is the wave length. The effect of the wave extends to a depth of half the wavelength, or from the example above, down to a depth of 125 feet—deep enough to alter the bottom structure of a considerable portion of the shelf.

The seafloor is also devoid of any major aquatic vegetation, which is common to the sounds, but this is not to say that the seabed is barren. There is a fantastic diversity of life, especially among the invertebrates such as worms, clams, and shrimp, many of which wind up in the stomachs of game fish. In particular, cobia are nicknamed "crab eaters" because the bulk of their diet is crabs. On a smaller scale, benthic algae and bacteria, single-cell microorganisms, are certainly the most numerous organisms on the seafloor. What they lack in size, they make up for in importance in the marine ecosystem. Bacteria perform a critical function: decomposing dead tissues, thereby recycling essential nutrients into the water for use by plants. Without decomposition, nutrients would remain chemically bound in the tissues of dead plants and animals. The result would be catastrophic, leading to a collapse of the entire ecosystem, since plants would not be able to photosynthesize without nutrients. No plants, no game fish.

The water column over the continental shelf, known as the neritic zone, is biologically linked to the seafloor below. Because of the sunlight available in the shallow waters combined with the abundance of nutrients (nitrates and phosphates) from bacterial decomposition, the neritic zone often teems with life. Just as spring signals the rebirth of life on land, the neritic zone experiences a virtual bloom of plants. Ranging in size from microscopic to tiny, phytoplankton are the major plant group. Collectively, phytoplankton are the "grasses" of the sea, providing nourishment for the entire pelagic food web. Since phytoplankton are the foundation of a food pyramid that supports apex predators, numerous pelagic species will migrate considerable distances, tracking the prey that is drawn to phytoplankton blooms. Zooplankton, which are very small planktonic animals, form the link between the phytoplankton that they consume and still other animals that eat these herbivores. The Atlantic menhaden, which spawns in the waters over the continental shelf and is a major food item for striped bass and bluefish, is an omnivorous filter feeder that strains food particles from the water. The species primarily consumes phytoplankton but also takes in a small portion of zooplankton. In the marine environment, the distinction between "Is that my meal?" and "Am I your meal?" becomes blurred to both prey and predator alike. But once this food chain becomes established in early spring, anglers take to the water in all types of watercraft to pursue their favorite species.

ACCESSING THE WATER

Though the maximum depth of the neritic zone is hundreds of feet, anglers frequently fish in relatively shallow water, often in sight of land, and on vessels ranging in size from an eight-foot kayak to a fifty-foot Carolina Sportfisherman. If you're not inclined to sitting an hour or more in a wet kayak, then by all means consider using a boat. For your own vessel, the National Park Service has boat-launching ramps near Oregon Inlet Fishing Center and Silver Lake on Ocracoke Island. On Hatteras Island, you will have to pay to use the private ramps. There is a relatively new launch ramp with plenty of trailer parking at Teach's Lair Marina. From this ramp, it's a relatively short run through Hatteras Inlet and out into the Atlantic.

Don't have your own personal watercraft? Many of the marinas offer inshore charters on either half-day or full-day schedules. These charters specialize in light-tackle fishing for mackerel (Spanish and king), trout, drum, bluefish, flounder, and cobia, to name a few. When sea and weather conditions are appropriate, the emphasis is on sight casting to visible and/or feeding fish. Other conditions may lend themselves to blind casting where certain species are known to congregate. The captain who operates a "for-hire boat" may have a "blanket" fishing license that precludes an angler from needing an individual license. Some may not, however, so it is a good idea to check beforehand. In addition, headboat fishing can give you a great fishing experience for a nominal cost ($30 to $70 per person). Headboats are built to accommodate a multitude of anglers (up to fifty); each person pays "by the head," thus the name. Plenty of open deck space from bow to stern allows anglers to stake out their own spot in comfort. All gear and bait are supplied, and a fishing license is not needed. If you are new to inshore fishing, the mates will help you with every aspect (from baiting your hook to identifying your catch) of this type of angling. Half-day trips stay close to shore and target small bottom fish, such as croaker, flounder, sea bass, and kingfish. In contrast, the full-day trips allow anglers to fish deeper waters that are farther from the shore. As with many other fishing opportunities, with the changing of the seasons comes a change in the species that anglers seek out.

TARGETING FISH: TACKLE AND TECHNIQUES

Spring means red drum, big ones. The ebb tidal deltas of both Hatteras and Ocracoke Inlets have red drum roaming the tops of these sandy shoals in search of prey. If the winds are relatively light, resulting in clear water, anglers will sight cast to schools of red drum. The primary method is to throw three- to five-ounce bucktails on medium-sized spinning outfits. If there is some discoloration to the water, adding a strip of bait to the bucktail takes advantage of a big drum's keen sense of smell. Fly anglers can also get their shot at these fish if they are capable of making long, quick casts using a nine-to ten-weight outfit with a fast sinking head. The key is getting the fly down to the fish that can be holding at depths over fifteen feet. Big flies, four to five inches long, which sink quickly and have the motion of a jig, round out the angler's arsenal. If the water visibility is poor, anglers switch to bait fishing, employing cut menhaden or mullet. Fishing these shoals is not for the

INSHORE FISHING

78

inexperienced boater, since even in moderate weather conditions, waves breaking over the shoals can lead to rough water that can damage a boat.

By early summer, when water temperatures creep toward 70°F, some hardy anglers will take to their kayaks and paddle out in search of Spanish mackerel. Though this is a small flotilla, it at times can be quite successful, since Spanish are notorious for staying just out of casting range of a shore-bound angler. The north and south sides of Cape Point are prime spots, with early morning and evening hours the most productive; however, if the water is clear, Spanish can be targeted throughout the day.

If you decide to try your luck from a kayak, you should make sure both your fishing tackle and safety gear (a personal flotation device is a must) are in order before leaving the beach. With regard to the former, a seven-foot spinning rod will allow you to cover the water, since this type of fishing is mainly blind casting to the points of a compass (N, E, S, and W). Spool the reel with six- to eight-pound test and tie on a twelve-pound fluorocarbon leader. Since you want to maximize your time on the water, take along an adequate supply of lures. If a cutoff occurs, and it invariably happens with Spanish or bluefish, you wouldn't miss the bite by having to paddle back to shore. I opt for small (one-half- to three-fourth-ounce) lures, such as Stingsilvers or Kastmasters; these sizes will often match the tiny baitfish (anchovies, silversides) that are in the water. Though space aboard a kayak is obviously limited, it is critical that you have some means of boating and restraining a thrashing fish, as well as removing the hook from the mackerel's formable dentition. Experienced kayakers, some offering fishing tours (see Appendix 4), have come up with some unique ways to efficiently rig their kayaks for fishing; you should check with them to get some ideas for customizing your kayak.

Trolling is the main method when targeting Spanish mackerel from a boat, be it small or big. Silver, gold, or painted lures (the Clarkspoon being a favorite) are trolled at about five to seven knots below the surface (ten to twenty feet) with the use of planers or in-line sinkers. For most shallow-water applications, size #1 or #2 planers and trolling weights between two and six ounces should perform well. Take along a variety of lure sizes to match the bait. Six-foot boat rods and conventional reels that are spooled with twenty-pound line will work just fine. The coupling link between the planer/sinker and the lure is a monofilament leader (twenty- to thirty-pound test) of twenty to thirty feet. But a longer leader may be needed if the water is particularly clear.

Of course, you can also catch bluefish and even king mackerel while trolling the inshore waters. If you're catching bluefish but not Spanish mackerel, you may want to increase your trolling speed a bit, since Spanish prefer the faster speed.

Some anglers modify their trolling setup by using a bird rig—a Clark-spoon (size #00 or #0) tied behind a small (five inches) plastic bird. The "flutter" that the bird makes as it skips over the water surface apparently imitates the sight and sound of an injured or panicked baitfish, which may prove irresistible to many fish. The rigging consists of the lure seven feet behind the bird on thirty-pound monofilament.

Throughout the summer, boaters routinely catch bluefish and Spanish mackerel. Boat one of these gluttonous predators, and it might regurgitate its last meal of silversides. Ever wonder why these fish have literally stuffed themselves with prey? Because of loss or utilization of energy along each step of the marine food chain, only a small percentage—10 to 15 percent—of the food consumed is converted to weight. This low efficiency does not bode well for blues or mackerels, which must replace weight lost during spawning or long periods of inactivity during the winter months. (A big bluefish caught during the spring will be long but have very little girth.) A 4-pound bluefish will have needed to eat approximately 40 pounds of baitfish to attain that size. But interconnections are quite common in the marine environment. It takes 400 pounds of planktonic animals to produce those baitfish, and a whopping 4,000 pounds of microscopic phytoplankton to produce the zooplankton.

During May and June, cobia begin to make an appearance, with the thickest concentrations often around Hatteras, Ocracoke, and Beaufort Inlets, as well as Cape Point and Cape Lookout. On calm days, when the water is clear, sight fishing is the method of choice. From boat towers, anglers have a good vantage point to spot cobia cruising just below the surface. It's a hunting game; boats run parallel to the beach, sometimes just within the outer bar-slough complex, and search for that tell-tale shadow. Cobia will often swim within the slough that they use as a conduit in their search for prey. Diamond Shoals can also be a hot spot for cobia because they often cruise the deeper periphery of this vast, shallow area. Upon sighting a cobia, the angler casts a bucktail, usually three to four ounces, in front of the fish. Quite often, controlled jigging or a slow retrieval of the lure will pique the cobia's interest. These can be big fish; stout rods and large spinning reels with fifty-pound braided line are needed to boat a fifty-plus-pound cobia. If sea conditions are a little rough, anglers may switch to anchoring, set out a chum bucket, and soak cut bait on the bottom.

In addition to targeting the summer staple of blues and mackerels, some adventurous anglers also try their luck in shark fishing for such species as duskys, hammerheads, and sandbars, to name a few. The general technique is set up over structure with large baits, one at the surface and one on a downrigger. Generally, it wouldn't be long before the angler is hooked up with a shark that might top at more than 100 pounds. While most marinas do not routinely offer shark-fishing charters, check around; some captains are willing to accommodate you.

Inshore fishing really heats up during the fall. September is a good month for flounder fishing in the inlets. Flounder are on the move, leaving their summer haunts to take up residence in the ocean during the colder months. While the migration may start in September, peak movement generally occurs in October and extends into early November. This time frame is particularly true for the southern inlets, such as Barden Inlet, the tiny opening between Shackleford Banks and Core Banks.

Drifting with the tidal current along the edges of an inlet's channel is the preferred method for catching flounder. This technique relies on the fact that flounder will typically face into the current, waiting for food to sweep past them. Seasoned flounder anglers are not in agreement as to whether an outgoing or incoming tide is more productive. For example, anglers fishing Beaufort Inlet, which has a high volume of water flowing through it because of its depth, generally prefer a falling tide to flush out the bait. Personally, I am of the opinion that the bite is, for the most part, independent of the tidal stage. Flounder are sedentary, ambush predators and will simply reorient themselves and face into the flow as the tide turns.

During the drift, many anglers employ a Carolina rig—a live-bait, Kahle-style hook on an eighteen-inch leader of twenty- to twenty-five-pound monofilament, with an egg sinker (generally one to two ounces) threaded above a barrel swivel. This setup will keep the bait just off the bottom—an ideal position for a flounder to swim up and grab it. Favorite baits include live menhaden or mullet minnows, in the three- to five-inch range. If these are not available, strip baits, such as squid, will work in the pinch. The underside (white flesh) of a legal flounder also makes a great strip bait, but if you clean any fish with a size limit, you must retain the carcass while onboard. While drifting, note where you catch a fish. Flounder are schooling fish, and it is advantageous to make repeated drifts over the area where you got that bite.

While tidal currents and even the wind may influence your fishing opportunities, be aware of their impact on your boating safety. Large rolling waves can develop at the inlet when tide and wind appose each other—for example, at Beaufort Inlet, a full ebb tide with a strong southeasterly wind.

Becoming more popular every year among anglers is the little tunny (false albacore) that is known for its line-burning runs and spirited fight on light tackle—medium spinning outfits to cast metal lures or nine- to ten-weight fly rods. These migratory visitors only make a relatively brief appearance during autumn, so if you're intent on catching one, you can significantly increase your chances, particularly if you are a novice, by hiring a guide who has experience in fishing for little tunny.

From mid-October through November, the waters near Cape Lookout, which are full of bay anchovies, a little tunny's main prey, are prime sites for this game fish. Boats leave Harker's Island and motor through Baden Inlet, a short ride for an opportunity to encounter schools of little tunny. These fish can be found anywhere between the surf line and a couple of miles offshore, depending on the concentration of bait.

The key to catching little tunny, whether using conventional or fly-fishing gear, is to cast quickly and accurately. These speedsters give anglers only a very brief window of opportunity to make the appropriately placed cast. Little tunny exhibit a relatively predictable pattern of behavior when feeding—rocking up to the surface to bust into schools of bait, scattering the bait, and diving deep again to wait for the bait to ball up again. It is critical to put the lure or fly into the boiling water before the fish sound again and move on.

King mackerel fishing can be quite good throughout the fall and into December. During these months, king mackerel can be found right off the beach, alleviating the need and expense of a long ocean ride. Inlets are quite productive because of the concentration of baitfish. Like many apex predators, kings are drawn to schools of bait; wherever there is a food source, there will be king mackerel.

A popular method for catching these fish is to pull multiple baits behind the boat at very slow speeds. Many anglers use live menhaden, but if they prove elusive to net, frozen bait (sardine minnows) can be substituted. Careful rigging of the live baits is essential for effective presentation and to keep

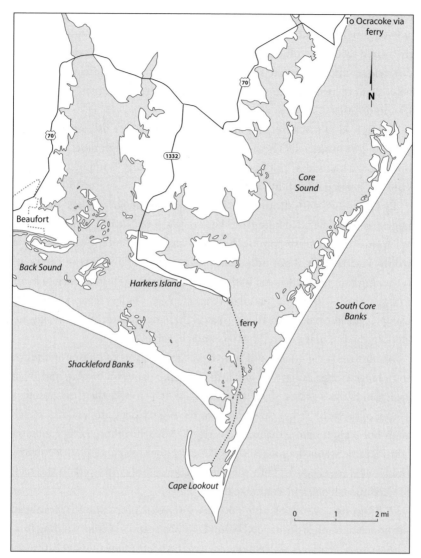

Cape Lookout

them frisky. Two small treble hooks make up the terminal tackle: one hook through the menhaden's nose and the other in the back near top center. The connective link between the hooks is generally a single-strand stainless steel wire; its small diameter minimizes water drag to allow the menhaden to swim naturally. The small hooks and light wire may also have the advantage of not spooking a wary king, especially in very clear water. But this down-sized rig may have a negative side if a big king inhales the offering, makes a long run, and subsequently pulls the hook. The trolling outfit is generally a conventional rod and reel, with line topping out at thirty pounds. To cover the water column, troll one flat line about thirty feet astern and one on a downrigger about fifteen to twenty feet in depth and twenty feet behind the boat. Want to avoid the hassle of catching and keeping live bait? Try trolling spoons or lures.

Though king mackerel run in the spring, the larger fish are almost always caught in the fall. These fish, which routinely weigh more than forty pounds, are known as "smokers" because they literally "smoke" a reel upon hookup, peeling line off in a blistering first run.

By December, with dropping water temperatures, striped bass will have migrated from their New England haunts south to North Carolina waters. With this influx of fish comes a hoard of anglers, all intent on catching a trophy rockfish. (It is not uncommon to see the parking lots of the boat-launching ramps overflowing with vehicles.) Without the presence of birds to signal the location of a school of stripers, many anglers resort to trolling, using either live bait, such as eels (if available), or an assortment of rigs, including the popular umbrella, Mojo, and parachute.

An umbrella rig consists of four to six metal arms (eighteen inches or so in length) that radiate from a hub (thus the name umbrella), and from each arm hangs a plastic shad (six inches). When trolled, the shads mimic a school of baitfish. To get down deep to the big stripers, anglers may attach an in-line weight before the umbrella rig. Because the umbrella rig generates considerable water drag when trolled, heavy boat rods and quality conventional reels are needed. (Thin wire line is sometimes employed on the reels to maximize depth and minimize drag.)

A Mojo rig consists of a big and heavy (two or more pounds) leadhead (an attached hook is positioned behind the head) and a smaller, trailing lure. The setup is relatively simple: one swivel from a three-way swivel attaches to the running line, a short leader from another swivel connects to the leadhead, and a longer leader from the remaining swivel connects to the lure, such as a swimming plug or small bucktail. The appeal of the leadhead is twofold: attaining deep depths and drawing strikes from big stripers. Most seasoned anglers pull multiple Mojo rigs at trolling speeds between three and four knots to cover the water column. However, depending on the boat, current, and nature of the rigs, experimenting with different trolling speeds may be in order. The key to success is positioning the rig near the bottom, but not constantly hitting bottom.

For anglers who do not like the extra weight and drag from an umbrella rig, a tandem parachute rig is a good alternative. A parachute is a bullet-shaped, leadhead lure (three to eight ounces) with a flared skirt (thus the name parachute), which increases the profile of the lure as it moves through the water. The main line is connected to one eye of a three-way swivel. From another eye, a leader of three to four feet connects a heavy parachute lure (eight ounces), and on the remaining eye, a leader of five to six feet connects a light lure (four ounces). Anglers can obtain the desired depth by altering the length of line out or the size of the lures in tandem.

With all these trolling options, you might be tempted to put out as many lines as possible. Many anglers believe the equation for trolling success is fairly simple: the more time they spend on the water, and the more lines they have in the water, the greater is the probability that they will attract the attention of a striper. While larger vessels, with an experienced crew, may be able to drag twelve or more lines through the water, anglers of smaller boats should be conservative in the number of trolling lines they employ, to avoid tangling of the various rigs.

Striper fishing can continue well into January, but its end signals a close to fishing the inshore waters until spring rolls around again.

THREAT TO INSHORE FISHING

Since 1993 a moratorium has been in place on oil and gas exploration off the North Carolina coast, thus protecting environmentally sensitive areas from drilling. However, over the years, Congress has periodically proposed various energy bills that contain provisions to open the coastal waters of North Carolina and other portions of the continental shelf to drilling. These proposals have met stiff bipartisan opposition, but will these areas still be off limits to oil companies in the near future as political pressure escalates to increase domestic oil production?

From a policy perspective, the Outer Continental Shelf (OCS) is considered federal waters, and the Minerals Management Service (MMS), a part of the Department of Interior, is charged with leasing and managing resources on the OCS. A particular site of interest for exploration is the Manteo Exploration Unit (MEU), an area about twenty-five miles wide off the North Carolina coast. Commonly known as "the Point" (not to be confused with Cape Point on Hatteras Island), the MEU is also known for its high biological diversity. The concern among exploration opponents is that the MMS would use tens of thousands of high-decibel explosives to gather vertical profiles of the seabed. While these seismic surveys are needed to reveal likely sources of oil and gas, the loud acoustic pulses are suspected of causing fish kills, whale strandings, and declines in fish catch.

While most oil rigs have a history of safe operation, the 2010 explosion of the drilling well Deepwater Horizon in the Gulf of Mexico should warrant a period of introspection on the consequences of a major oil spill. Crude oil is a complex mixture of hydrocarbons, some chemically light in composition and others heavy. Thus the oil enters many pathways into the marine environment: dissolving in seawater, forming large surface slicks, and sinking to the bottom.

Reef, Ledge, & Wreck Fishing

Offshore bottom-fishing depends on locating fish structure, such as ship-wrecks, artificial reefs, rocky outcrops, and ledges. As we will soon see, many of these sites are documented, but others require systematic searching by the angler.

The treacherous waters off the Outer Banks are known to both mariners and historians as the last resting place for all types of vessels. Since 1585, when Sir Richard Grenville's flagship, *Tyger*, ran aground near Ocracoke, the sea has claimed hundreds of vessels. A perfect storm of geography and weather combines to create a deadly situation for vessels traveling along the coast. Diamond Shoals, which extends twenty miles out to sea, is only a few feet deep in places and, even today, is essentially uncharted. Add the common occurrence of nor'easters and hurricanes, and you have a recipe for maritime disasters. As might be expected, the majority of these shipwrecks lie in the relatively shallow waters (50 to 200 feet) of the continental shelf. The Association of Underwater Explorers maintains a website that provides detailed maps and locations of numerous shipwrecks off the North Carolina coast (http://uwex.us/northcarolinashipwrecks.htm).

The State of North Carolina has vigorously attempted to improve fish habitat by installing a number of artificial reefs on the continental shelf. The North Carolina Division of Marine Fisheries pinpoints the location of a number of these reefs on its website (www.ncdmf.net). These reefs are com-posed of a variety of objects: concrete shells, railroad cars, bridge rubble, barges, and even a Falcon 150 aircraft. Many of these structures are only a few miles from the buoys that mark the entrances to the inlets and thus are readily accessible even by small boats.

Since most of the above sites are well known to both recreational and commercial anglers alike, they can receive considerable fishing pressure over time. Charter boat captains who come upon a new productive bottom structure are naturally reluctant to divulge its location. However, anglers can score on their own by using their depth finder to pinpoint fish-holding structures, such as rocky outcrops and sharp drop-offs on the continental

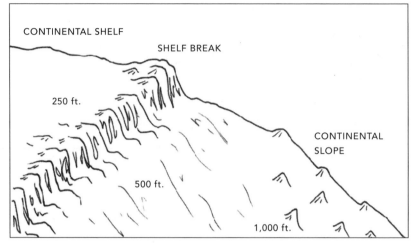

CONTINENTAL SHELF

SHELF BREAK

250 ft.

CONTINENTAL
SLOPE

500 ft.

1,000 ft.

Continental shelf and continental slope

shelf. These spots may be small, but they can offer suitable habitat for big fish.

When the weather permits, boats make the run to the deeper fishing grounds that may be thirty to fifty miles from shore. These sites occur at the shelf break—a bathymetric transition from the relatively flat, shallow continental shelf to the steeper and deeper continental slope. At the break, the seafloor plummets from 260 feet to more than 600 feet. Undersea slides and slumps have formed holes, ledges, boulder fields, and depressions that attract a variety of bottom fish, including tilefish (both blueline and golden), snowy grouper, and red porgy. Sites quite familiar to both commercial and recreational fishermen are the Outer Shelf Reefs, which as the name suggests are located at the edge of the shelf. These reefs, located between 160 and 650 feet in depth, are covered with sponges, soft and hard corals, bivalves, and hydroid colonies—all providing important fish habitat. Researchers from the University of North Carolina at Wilmington found robust populations of sea bass, vermillion snapper, grunts, and snowy groupers.

An added factor in fish aggregation is the position of the Gulf Stream. When the western wall of this current lies directly over the shelf break, this can lead to extraordinary bottom-fishing for grouper and snapper. During early spring, when the coastal waters are quite chilly, the warm edge of the Gulf Stream draws all kinds of bottom fish and concentrates them into a smaller area. In addition, with the presence of a southerly wind, surface waters become nutrient-enriched due to a phenomenon known as shelf break upwelling. Nutrients, which support plant growth, are brought up from the deeper continental slope and set the stage for a biologically productive shelf break environment.

On the other hand, when warm core rings that spin off from the Gulf Stream interact with the continental shelf, unique circumstances develop that can have enormous practical effects on the biology of the area. And the most visible effect of the warm core rings is their impact on fishing. When a ring interacts with the shelf, it creates turbulence and shearing at the level of the shelf because the ring's depth is much greater than the shelf's. As a result, water from the ring is transported onto the shelf, and the shelf water is pulled into deeper water.

This transport has a potentially negative impact on the survival and recruitment of many of the main food fish in the area. In particular, shock of the larvae results from the sudden increase in temperature associated with this water exchange. While this shock may not directly kill the larvae, their growth is often impaired. Recruitment is affected when the larvae or young fish are swept off the shelf into deeper water. When ground fish mature enough to make their trip to the bottom, their evolutionary imprint doesn't include a guarantee that they are over shallow water. In effect, the fish simply try to get to a bottom that is beyond their ability to reach; thus they effectively remove themselves from future fish populations. On the other hand, researchers have demonstrated that bluefish recruitment to Middle Atlantic Bight (a coastal region running from Massachusetts to Cape Hatteras) estuaries increases with increased warm core ring activity. In summary, warm core ring activity has been associated with both increased and decreased fish recruitment, depending on the species and the locations of the rings on the shelf.

BEFORE THE FISHING

Bottom-fishing, whether in shallow or deep water, has a broad appeal because it offers a wide range of game fish, from snapper and grouper to black sea bass and amberjack, that cling tightly to their piece of the bottom. Since these fish are highly territorial, success depends on first finding the wreck or reef and then positioning the boat as close to the structure as possible.

Even with Global Positioning System (GPS) coordinates, pinpointing a particular structure to fish is often the most difficult and time-consuming part of this type of fishing. When you are within a hundred yards or so of your target, slow down and systematically begin prospecting the area with your depth finder. If nothing encouraging shows up on the screen, proceed to the GPS waypoint and drop a marker buoy (an old crab pot buoy or plastic jug works well) overboard on the exact spot. With that as your reference point, make ever-widening circles around the buoy. On the depth finder, you are looking for abrupt changes in depth and/or changes in bottom hardness.

Upon locating the wreck or reef, spend some additional time making a few more passes over it. You want to be able to answer the following questions about the structure: How is it oriented? What is its overall size? Where in relation to the structure are fish marks showing up on your screen?

When fishing the relatively shallow wrecks, reefs, and rock piles, anchoring, rather than drifting, is the preferred technique. While the Danforth anchor is employed for digging into and holding in sandy and muddy bottoms, savvy bottom-anglers use a grapnel—or grappling hook—which consists of a metal cylindrical shank terminating in three to six flexible, clawlike prongs that snag a piece of the bottom. At the day's end, anglers can pull free the anchor from the structure by tying it off on a boat's cleat and motoring away from the snag to straighten out the pliable arms.

If you are hesitant to anchor directly over a wreck, try anchoring up-current of the structure, so that your offering can drift over the target area. In most cases, it is better to anchor farther from the structure than you initially deemed necessary; you can always let out more line to drift into the "proper" position. Though simple in principle, this technique can be quite tricky. Changes in current and wind may cause the boat to not ride in the same place for any extended period of time.

Be a courteous angler and follow some simple etiquette guidelines when fishing structure: (1) Avoid anchoring on structure when other boats are already drifting. Your option should be to join the drift pattern, but up-drift so as not to impede the progress of another angler's boat. (2) If other anglers are chumming, then by all means do not anchor in the chum slick. Leave plenty of room between you and the closest boat to allow for swing on the anchor. (3) Do not attempt to fish an already crowded small structure. (4) A section of the structure belongs to the first person fishing it.

Before attempting to fish the deep waters of the shelf break, your first task should be to assess if weather and sea conditions are favorable for the long run to these fishing grounds. It can be quite discouraging after navigating miles of open ocean to encounter a strong current that limits your fishing opportunities. Keep in mind that the Gulf Stream, which brushes up against the shelf edge, changes seasonally in both position and strength. On the average, this current shifts northward and increases in strength during the fall and moves southward and weakens during the spring. Maps that show the daily position of the Gulf Stream as well as its velocity can be an invaluable aid to both novice and experienced anglers. Rutgers University's marine science program produces daily sea surface temperature maps from satellite data that are free to the public (www.marine.rutgers.edu/mrs/sat_data).

If you decide to make the trip, take along a good nautical chart of the offshore waters that allows you to locate the edge of the shelf. Using the chart and your depth finder, carefully search the area for marked changes in bottom relief—hills, humps, ridges, or ledges. At these great depths, anchoring is not possible, so fishing is done on the drift and using electronics, like GPS, to return to the structure. The key is to drift the whole structure and slightly beyond it, so that "hot spots" on the structure can be pinpointed. A light current is ideal, since the angler can stay in contact with the bottom. A slow and controlled drift over the structure allows the baits to continually drift over new bottom and may even trigger a lethargic fish, not wanting to miss an easy meal, to leave its lair and strike the moving bait.

Even a slow drift can be quite challenging, since the angler must drop the bait from an extreme depth inside a small chasm or hole. If the wind or current kicks up and the boat drifts too fast, the drop's accuracy is severely compromised. In this case, the captain must power the boat over the mark and keep it in position. This tactic, while effective, can be expensive, burning a considerable amount of fuel.

TARGETING FISH: TACKLE AND TECHNIQUES

When fishing for most structure-oriented bottom fish, such as black sea bass, blueline tilefish, groupers, and snappers, the two factors that determine your selection of tackle are the depth of the water and the size of the species.

Fishing the shelf break and beyond, often called deep-dropping, requires heavy weights. A general rule is one pound of lead for every 100 feet of depth. You want your line straight down. With a lighter weight, a bow will form in your line even under moderate current conditions, and your rig will never reach the bottom. Your rig's trip to the bottom can take a minute or longer, even with a few pounds of weight. When the weight hits the bottom, you want to keep it bouncing off the bottom and straight down under the boat.

Braided line (50- to 100-pound test) is also a necessity because its small diameter decreases water friction. In addition, braided line, which has minimal stretch, allows extra sensitivity to detect strikes and minimizes current-induced bowing of the line. (On shallower wrecks and reefs, anglers can substitute monofilament [30-pound test] for braided line and use less weight.)

Deep-dropping necessitates the use of a heavy rod and reel combo. Because of the extreme depths, electric reels are the sensible choice. A high line capacity, 80-pound-class electric reel is coupled with an 80- to 130-pound heavy action rod. While pressing a button may be a simple way to reel in a prize fish from the depths, be aware that the use of an electric reel may disqualify the angler for any subsequent fish records. If deep-dropping interests you, then you might want to try a charter first because these rod and reel outfits are expensive—$2,000 to $3,000.

Anglers can get by with conventional boat rods (30- to 50-pound class) when fishing shallower water (200 to 300 feet), but they should be prepared for a strenuous workout. Even without a fish on the line, you will have to do a great deal of hard and long cranking to get the weight up to the surface. Some charter boat captains employ the same rods they use to target medium-sized dolphin.

The terminal tackle for many of these species is relatively simple and can be used for both shallow and deep bottom-fishing. For wrecks or reefs up to 150 feet in depth, use a 50-pound monofilament leader, tie in a couple of dropper loops, attach a couple of hooks, and connect a sinker to the end of the leader. The sinker can be tied using a relatively weak knot so that if the sinker gets hung up on the structure, the rest of the rig can be saved.

Targeting heftier fish calls for beefing up the tackle a bit: to the main line, attach a three-way swivel, connect two feet of 100-pound leader material from a swivel to the hook, and connect two to three feet of the leader from the remaining swivel to the sinker. Some anglers even use a heavier leader so that multiple hooks can be used on the rig. Hooks can include either 6/0-sized J hooks or 10/0 to 12/0 circle hooks. Since deep-drop fishing does not entail setting the hook in the traditional sense, circle hooks may improve catch numbers.

Many of these fish are not too selective with regard to what they will eat, so almost any cut bait—squid, clam, and fish—will work just fine. Fresh bait is preferred over frozen because fresh is tough enough to stay on the hook; frozen bait is too soft when it thaws. Make sure the bait is securely impaled on the hook for its long ride down to the bottom.

After the drop, although the rod is in the holder, the angler should monitor the line to ensure the rig is on the bottom and to watch for the telltale twitch of the rod tip, signaling a fish has engulfed the bait. When the tip bounces—braided line telegraphs that quite nicely—quickly throw the electric reel into gear.

If done correctly, deep-drop fishing can yield surprisingly good results. At these great depths, the amount of food available to resident fish is quite scarce, so they will readily inhale any natural bait that you offer them. And the fish can be big. Smaller fish are generally scarce, but if they are present, you can deter them by using large, tough strip baits. These small fish can pick at but not swallow the bait.

Many deep-dwelling fish have a gas-filled (mostly oxygen, carbon dioxide, and nitrogen) organ known as a swim bladder, which controls buoyancy and allows them to maintain a certain depth in the water column. When a fish is reeled quickly to the surface from the depths, the gases in the bladder expand due to the decrease in water pressure. Visible signs of this affliction include protrusion of the stomach from the mouth, bulging eyes, and a bloated stomach. If the fish is released in this buoyant condition, it cannot return to the bottom but may simply float away and die from exposure or predation. This defeats the purpose of fishery management regulations, such as minimum size restrictions and daily bag limits. For the fish to survive and avoid permanent injury, anglers must release or vent the trapped gases. Though there is some debate within the angling community about the benefit of the venting procedure, a number of scientific studies have shown that species with large swim bladders, such as red grouper, black sea bass, and red snapper, derive almost immediate relief from venting.

Deflation of a bloated fish is accomplished with a venting tool—any hollow, sharpened instrument, such as a hypodermic needle, that allows gases to escape. (Ice picks and knives are not suitable, since simply puncturing the fish can result in its death. Venting tools are also available from a number of fish-tackle supply outlets.) The venting should be done quickly and with as little handling of the fish as possible. While I won't go into great detail about the procedure (take a look at some of the videos on the Internet that show it), venting involves inserting the needle into the body cavity at a forty-five-degree angle under a scale in an area approximately one to two inches behind the base of the pectoral fin. If the fish is vented properly, you will hear the gases escaping. Then release the fish back into the water, making sure that it is strong enough to swim away under its own power.

THREATS TO STRUCTURE FISHING

The biology and behavior of these fish make them susceptible to overfishing. Congregating in relatively small sites, populations can be severely depleted due to increasing fishing pressure and can take decades to rebound. In addition, many of these species are slow growers and reach maturity at a relatively old age. Anglers targeting these big spawners can quickly deplete the stock. Heed the well-worn maxim: limit your catch and not catch your limit.

The South Atlantic Fishery Management Council (SAFMC), which monitors and oversees the health of North Carolina's fisheries between 3 and 200 miles, has determined that overfishing of snappers and groupers is a problem on the Outer Shelf Reefs. Since neither species recovers quickly after the cessation of fishing pressure, the SAFMC may consider designating this area and potentially other threatened sites as Marine Protected Areas (MPAs) or Habitat Areas of Particular Concern (HAPCs). MPAs are areas of the marine environment where natural and cultural resources enjoy greater protection than the surrounding area. Depending on the degree of protection, commercial and recreational uses may vary. At present, SAFMC has designated eight MPAs within the South Atlantic region, one off southern North Carolina (Snowy Wreck). HAPCs are threatened water columns or seafloor habitats recognized as critical to the health and long-term viability of fisheries.

Fish Anatomy *An Overview*

In Chapters 9 and 10, you will become familiar with the major game fish that swim the nearshore and offshore waters of the Outer Banks, from the sedentary bottom-dwellers to the highly mobile pelagic species. Their morphological and physiological characteristics allow them to adapt to and survive in their particular habitat. From an angler's perspective, a rudimentary understanding of some of these characteristics will allow you to better know your quarry, from its swimming style to its detection of prey to its feeding strategy.

For many sportfishermen, members of the Scombridae family, which includes tuna, mackerel, and little tunny, are prime targets. While all members exhibit similar traits, tuna may represent the pinnacle of marine engineering by Mother Nature. Tuna have a highly streamlined body—fusiform shape—tapering at both ends from a thicker midsection. Endowed with this hydrodynamically efficient body, which results in a smooth, uniform flow of water over its surface, thereby reducing drag, the tuna swims effortlessly through its fluid environment, overtaking its prey with relative ease. Rudderlike finlets along the top and bottom of the caudal peduncle (the section immediately anterior to the fish's tail) aid in the tuna's locomotion. As the fish moves through the water, these finlets generate little vortices, similar to dimples on a golf ball in flight, which contribute to the reduction of a turbulent wake.

Though the tuna is blessed with a streamlined body, propulsion is a complex interaction between the tuna's muscles and its caudal fin (tail). As an apex cruiser of the pelagic environment, a tuna has a lunate (sickle-shaped) caudal fin, which has a minimum surface area but has maximum spread between the top and bottom lobes of the fin. Biologists have determined that these dimensions minimize speed-sapping drag, resulting in sustained motion. This very rigid fin, slicing through the water at an eye-blurring twenty beats per second, exerts a significant back pressure that provides the forward thrust of the fish. But specialization in one direction requires trade-offs in other areas. Though a tuna can cover large stretches of the watery realm, it has limited turning ability. Because of this lack of maneuverability, a tuna often fails to capture a significant number of baitfish from the schools it encounters.

Bluefin tuna (courtesy of NOAA)

In contrast, demersal fish, such as flounder, which live near or on the sea-floor, employ both camouflage and stealth to ambush their unsuspecting prey. Possessing a depressed (vertically flattened) body shape, flounder can lie flat and motionless on a sandy or muddy bottom. (In contrast, eels have an elongated body.) When opportunity presents itself, they quickly accelerate upward, powered by their broad tail, to engulf their prey. But these sedentary fish could not adapt to the open waters of the Gulf Stream because the downside of a broad tail is increased drag, resulting in loss of sustained speed.

Before fish pursue their prey, they must rely on their senses to detect their next meal, with some fish depending more on certain senses than others. Fish, such as spot, that forage near the bottom or in generally discolored water have a well-developed sense of smell. Their olfactory system consists of nostril-like holes called nares. Four nares are located close to the top of the snout, two on each side. These openings don't lead directly to the throat, the way nostrils do in mammals, but open into nasal sacs that contain complex chemoreceptor cells. Water-borne odor is drawn into the sacs and passes over receptors of smell, which are connected to the brain by means of olfactory nerves.

Flounder and eel (courtesy of NOAA)

Not all fish move water in and out through these nares in quite the same manner, but the key to possessing a strong sense of smell is the ability to move water rapidly over the sensory parts. Some fish can pick up chemical signals when immobile by pumping water through their olfactory system. Other fish, such as mackerel and tuna, have an olfactory system that requires them to swim, forcing water through their nares.

In contrast, sight-dependent predators are ideally suited to the clear waters of the open ocean and Gulf Stream. The large eyes and well-developed visual centers in the brains of tuna and marlin (almost 30 percent of a marlin's brain is devoted to analyzing the visual signals coming from the eye) strongly suggest that vision is an important sense for feeding and other behaviors. But pelagic species differ in their ability to detect objects in the varying light conditions of the open ocean. A dolphin mainly forages within the upper, sunlit layers of the Gulf Stream and primarily during daylight hours. In contrast, the optical sensitivity of the marlin's larger eye is higher because of the large number of light receptors it contains. The marlin's eye is specifically adapted to cope with low light levels encountered during a dive, so the fish's feeding range is increased.

Fish have evolved to have different types of mouths, depending on what their diet consists of and how they feed. For example, mackerel and tuna, which feed on other fish in the water column, have a terminal mouth at the end of the head. In contrast, a fish possessing an inferior (underslung) mouth—a mouth located on the underside of the head and that opens downward—generally signals that the fish is a bottom-feeder.

Within the fish that roam the waters of the Outer Banks, a large variation occurs in the shape and size of their teeth. The nature of the dentition is an angler's clue to the fish's feeding habits and the type of prey it consumes. Many piscivorous predators, with wahoo probably topping the list, have an awesome array of teeth—flat, triangular, closely set, and extremely sharp. These teeth are ideally suited for capturing live fish and cutting larger victims in half. Bluefish, well known in angling circles for their reputation of chopping up a school of baitfish, have similar but smaller teeth. Other species, spotted seatrout and gray trout being two good examples, have sharp, conical teeth (also known as canine teeth) that cannot cut but perform quite well in piercing and grasping the prey until its struggles diminish. Some bottom-dwelling species, such as drum, have molarlike teeth that are well adapted for crunching mollusks and crustaceans.

Spot with underslung
mouth (author's photo)

Wahoo (author's photo)

Skipjack tuna
with terminal mouth
(author's photo)

Inshore &
Coastal Species

BASS, STRIPED (*MORONE SAXATILIS*), ROCKFISH, STRIPERS

N.C. Citation Size: Thirty-five pounds or live release of fish forty-five inches or longer.

Availability: November–January.

Coloration: Olive green, blue, or black on top, silvery body with lateral dark stripes (seven or eight) stretching from behind the gills to the base of the tail, and white on the underside.

Distinctive Features: Mature striped bass have an elongated, stout body with a long head, pointed snout, and large mouth. The caudal fin is broad and forked. Adult stripers are known for their size, reaching nearly five feet in length and routinely over fifty pounds. Presently, the all-tackle record for the species is seventy-eight pounds, eight ounces.

Diet: Larval striped bass feed on zooplankton, primarily copepods (tiny crustaceans); juvenile stripers eat insect larvae, larval fish, and mysids (shrimplike crustaceans); adult bass are piscivorous (fish-eaters) and eat almost any kind of small fish: bay anchovy, menhaden, herring, spot, and croaker.

Biology & Behavior: Striped bass are anadromous fish, meaning they migrate from salt water to fresh water to spawn. Major spawning sites include the rivers and tributaries of the Chesapeake Bay and the Hudson, Delaware, and Roanoke Rivers. Many of the local North Carolina striped bass caught at Manns Harbor or Oregon Inlet are likely from the Roanoke River population. Spawning behavior frequently centers on a single large female being courted simultaneously by a number of smaller males. Shortly after spawning, mature fish return to the coast. Most stripers spend the summer and early fall in New England waters, and by late fall and early winter they migrate south off the Virginia and North Carolina capes.

The striped bass's common regional name, rockfish, derives from its frequent association with rocks, and its species name, *saxatilis*, means "rock dweller." But along the Outer Banks, stripers are to be found around piers, groins, surf sloughs, and rips.

Angling Techniques: The level of striper activity determines the fishing method. In the infrequent case of striped bass actively chasing bait at or near the surface, casting lures into the melee will almost guarantee a hookup on every cast. Though each striper aficionado has a favorite lure, many will have on hand a couple of top-water and darting plugs as well as a few long-casting metal spoons that will allow them to cover the water column and reach distant stripers. If there are no visible signs of stripers feeding, then a bucktail fished around a jetty (Oregon Inlet) or a groin can be particularly effective because of a striper's affinity for structure.

Outer Banks surf anglers also catch striped bass by fishing fresh, cut bait (menhaden, mullet) on the bottom with a nine- to ten-foot spinning or conventional outfit. This technique depends on the water temperature being within the stripers' preferred range (50° to 60°F) and the presence of natural bait in abundance. Bottom-fishing for big stripers is generally better at night because they are more comfortable feeding in the shallows under the cover of darkness.

Anglers fishing the inlets and inshore waters often troll at relatively slow speeds (two to three knots) a variety of rigs: Mann Stretch lures, umbrella rigs, and parachutes, to name a few.

Angling Tip: At Oregon Inlet, some anglers use a live eel that is cast out and slowly retrieved back to shore.

Table Fare: Excellent. A happy balance between meaty and flaky.

BLUEFISH (*POMATOMUS SALTATRIX*)

N.C. Citation Size: Fifteen pounds.

Availability: April–May and October–November (jumbo bluefish), May–September (smaller bluefish).

Coloration: Bluefish are grayish blue-green dorsally and fade to white on the lower sides and belly. A dark blotch is found at the base of the pectoral fin.

Distinctive Features: The bluefish has an elongated, moderately compressed body that ends in a broad, forked tail. The upper and lower jaws of its rather large mouth have a single row of razor-sharp, triangular teeth.

Diet: The bluefish is a pelagic fish that is considered a bloodthirsty predator of other fish, including menhaden (a favorite), silversides, weakfish, grunts, and jacks. It has the dubious reputation among marine predators of wantonly killing prey that it does not consume. During feeding frenzies, commonly known as blitzes, bluefish have literally driven schools of prey onto a dry beach.

Biology & Behavior: Being cannibalistic, bluefish tend to school and migrate according to size: snappers (less than a pound), tailors (two to five pounds), and choppers (more than ten pounds). These aggregations can cover tens of square miles. Along the eastern seaboard, there are at least two groups of bluefish that are from two distinct spawning congregations, separated by Cape Hatteras. The Gulf Stream transports larvae spawned south of Cape Hatteras northward, where eddies can spin off and carry the larvae into the mid-Atlantic group. Historically, the bluefish population has been highly cyclical, and abundance has oscillated

over a span of a decade or more. During the 1970s and 1980s, large
schools of jumbo bluefish would routinely show up in the coastal
waters of the Outer Banks during the fall, but over the years their
numbers have decreased markedly. A large bluefish may live for
more than twelve years.

The snapper blues are often found in bays and estuaries. In contrast, the adults migrate along the coast—northward in the spring
and southward in the fall—and are caught by both shore-bound
anglers and boaters.

Angling Techniques: To many surf anglers, bluefish are the quintessential coastal game fish that are known for their tenacious fighting
ability and their eagerness to take a wide variety of bait and lures.
Menhaden, mullet, squid, shrimp, or similar bait are all productive, particularly when they match the natural prey that bluefish
are feeding on. Shiny lures (Hopkins, Stingsilvers), bright-colored
plugs, jigs, and bucktails all take their fair share of bluefish. Noisy
surface lures, such as top-water poppers, can be effective when
bluefish are chasing prey on the surface during dawn and dusk.
Light to medium spinning or bait-casting outfits (eight- to ten-
pound test line) are suitable for targeting smaller bluefish. For
choppers, stouter rods are needed, and they are matched with
reels that hold seventeen- to twenty-pound test line.

Boaters catch their share of bluefish by trolling small lures, such
as Clarkspoons, at a speed of five to six knots. Planers are generally
employed to position the lures at a depth that the fish are holding.

Angling Tip: Steel leaders offer protection from the bluefish's sharp
teeth, which can cleanly snip through monofilament line.

Table Fare: Good. Smaller fish (two to three pounds) are preferable to
larger ones. In all cases, the center bloodline should be removed
to improve palatability.

COBIA (*RACHYCENTRON CANADUM*)

N.C. Citation Size: Forty pounds or live release of fish thirty-three inches or longer.

Availability: Late May to early July.

Coloration: The adult body is dark brown with a white underside and has two narrow dark bands (silver or bronze) extending from the snout to the base of the caudal fin. The locals often refer to this fish as the "man in the brown suit."

Distinctive Features: A cobia has an elongated, torpedo-shaped body with a long, depressed head. A cobia's large pectoral fins are oriented in a horizontal, winglike fashion to the body, so that when seen in the water a cobia is often mistaken for a shark. The skin is smooth with tiny embedded scales, and the first dorsal fin is composed of seven to nine short but sharp spines. The genus name *Rachycentron* comes from the Greek *rhachis*, meaning "spine," and *kentron*, meaning "sting." The caudal fin is round to truncated in young fish, but as maturity occurs, the fin assumes a forked appearance.

Diet: Cobia are voracious eaters, often engulfing their prey whole and feeding on crustaceans (portunid crabs being a favorite), cephalopods (octopus), and small fish (pinfish, grunts, croakers).

Biology & Behavior: Researchers believe that cobia spawn in offshore waters from April to September. The resulting drifting larvae grow quickly, and cobia mature within two to three years of their birth, ultimately achieving life spans of fifteen years. There are two distinct stocks of cobia—one in the Gulf of Mexico and another in the waters along the southeastern coast of the United

States—and the federal agencies responsible for their management assume no mixing between the two groups. The species is highly migratory, making seasonal journeys along the Gulf and Atlantic coasts. After wintering in the Florida Keys, the Atlantic stock migrates north as far as Maryland in the summer, passing Florida's central coast in March. Along the North Carolina coast, cobia form large spawning aggregations during May and June. They will often cruise the shallow waters along the shoreline and forage in schools ranging in size from a few fish to more than dozens. Cobia are often seen in the company of cruising rays and may scavenge whatever the rays stir up from the bottom.

Angling Techniques: Cobia are powerful, hard-fighting fish that demand a sturdy rod and reel combination whether you are fishing from beach, pier, or boat. For example, surf anglers opt to use a ten- to twelve-foot rod, spinning or conventional reels with a smooth and well-maintained drag to put the brakes on a cobia's inevitable long run, and fifteen- to twenty-pound line with a leader doubling the line size. From the beach, cobia fishing is a waiting game, soaking a hunk of cut bait (menhaden, mullet) on a large (6/0 to 9/0) circle hook. In contrast, boaters sight cast jigs (three to six ounces) to migrating fish. Cobia appear to prefer clear water, and most anglers will target these species only if the water clarity is optimum. Many shore-bound anglers will often carry the bait out to clearer water by means of a kayak.

Angling Tip: When boated, the horizontal pectoral fins enable the cobia to remain upright, so its vigorous thrashing can cause damage to the boat and injury to personnel aboard. Anglers should immediately control the fish and place it in the fish box.

Table Fare: Excellent. The firm, white flesh is amenable to many forms of cooking.

CROAKER (*MICROPOGONIAS UNDULATUS*)

N.C. Citation Size: Three pounds.

Availability: May–August.

Coloration: Adults are silver with a pinkish hue, but older fish are brassy in color, with numerous brown spots that form oblique wavy bars on their backs.

Distinctive Features: The croaker has a short, elongated body with a short, high first dorsal fin and a long, low second dorsal fin. The croaker's chin has three to five pairs of small barbels or "whiskers" that help it feel for food on the bottom.

Diet: Since the croaker's mouth is located at the bottom of its head, facing the ground, this fish is an opportunistic bottom-feeder, consuming a variety of invertebrates.

Biology & Behavior: Croakers belong to the family of fish called Sciaenidae, which includes spot, black drum, and red drum. This species gets its name from the deep, croaking sounds generated when it vibrates its swim bladder through muscular contractions. This behavior is part of the male's courting ritual to attract females.

During the fall spawning season, females will release upward of 2 million eggs. Upon hatching, the larvae take up residence in shallow coastal waters with soft bottoms. But predation is high—more than 95 percent of the Atlantic croaker larval and juvenile population dies each year—from striped bass, spotted seatrout, other croakers, and humans. Those that survive can live up to eight years.

Though the juveniles are estuarine-dependent, croakers become oceanic during spawning. Their choice of habitat is eclectic, including mud, sand, and shell bottoms. Adult croakers can tolerate large ranges in temperature (50° to 90°F) and large variations in salinity (twenty to seventy parts per thousand).

Angling Techniques: For surf anglers, an eight- to nine-foot spinning rod and reel combination, designed for ten- to fifteen-pound line, is appropriate. A two-hook bottom rig, with enough weight (two- to three-ounce pyramid sinkers) to hold bottom, completes the setup. Baits include pieces of shrimp, bloodworms, and squid. The relatively new synthetic baits, such as Fishbites, have proven to be a viable alternative to natural baits, catching their fair share of fish. Whatever the offering, anglers should seek out suitable fish-holding structure, such as sloughs. Anglers fishing from a boat or pier may use the same bottom rig as surf casters but may want to slowly lift and lower their rod tip to provide movement to their offering. In addition, drifting, as opposed to anchoring, allows boaters to cover more territory, increasing their chances of coming upon productive bottom structure.

Angling Tip: Bottom-feeders such as croakers appear to cooperate best when the water is discolored with wave-stirred bottom sediments.

Table Fare: Good. When fried, the white flesh is mild but with a slight underlying sweetness to it.

DRUM, BLACK (*POGONIAS CROMIS*)

N.C. Citation Size: Thirty-five pounds or live release of fish forty
  inches or longer.

Availability: April–July.

Coloration: The body color in adults is gray to black, generally highlighted with a brassy luster. The fins are dusky to black. In contrast, younger fish have four to five dark vertical stripes, so the
  angler might mistake younger black drum for sheepshead at first
  glance. These stripes usually disappear with age as the fish grow
  from twelve to twenty-four inches in length.

Distinctive Features: The black drum is a chunky, high-backed fish
  with a short head tapering to a blunt snout. The chin has about
  a dozen short barbels that are set close to the inner edges of the
  lower jaw. The dorsal fin is continuous, but a deep notch separates the spiny anterior portion of the fin from the soft posterior
  portion. The teeth are small and set in broad bands for effective
  grinding.

Diet: Black drum are primarily bottom-feeders. The juveniles feed
  on worms, small shrimp, and soft crustaceans, but large drum consume mostly mollusks and crabs. The fish employ their chin bar-
  bells to search out food.

Biology & Behavior: This fish is a member of the Sciaenidae family,
  which includes the Atlantic croaker, spotted seatrout, and red

drum. Black drum are the largest of the aforementioned species, reaching weights in excess of ninety pounds. The current world record tops out at just over 113 pounds. Similar to its croaker cousin, the drum gets its name from the drumlike noise it emits by modulating its swim bladder.

Though detailed growth information for black drum is relatively scarce, there is no evidence of sex-specific difference in growth rates. Large specimens weighing over forty pounds can be either male or female. Males reach sexual maturity at a slightly younger age (four to five years) than females (five to six years), and spawning occurs in the coastal bays and estuaries, where the fish use the sea grasses as nursery habitat. These fish are prolific spawners. A relatively recent study estimated fecundity of average-sized females weighing approximately fourteen pounds at 32 million eggs annually.

Black drum can adapt to a wide range of habitats, including the clear waters of sand flats, the turbid waters of an estuary, and oyster reefs. In shallow, muddy waters, drum often dig or root out buried mollusks and worms, creating small depressions in the soft bottom.

Angling Techniques: Since black drum are solely bottom-feeders, the technique is relatively simple: fish the bottom with a baited hook, sit back, and wait for the drum to take your offering. Arguably, the baits of choice are clams and blue crabs. Depending on the size of the crab, impale a whole or half crab on a relatively large hook (8/0), particularly when targeting big drum over forty pounds. Clams need to be tied directly to the hook; a twist tie works well here. Match the tackle to the size of the drum. Small drum (five to ten pounds) can be quite sporting on light spinning tackle, but the angler should upgrade to stout rods and fifty-pound test line when tackling the bigger brutes. While these fish wouldn't make any longs runs, be prepared for a hard and slow fight. A rhythmical pumping of the rod will allow you to retrieve line and dislodge the fish from the bottom. Knowledgeable anglers target areas of moderate tidal flow, such as inlets, and/or sites with structure, such as jetties.

Angling Tip: Drum will often "mouth" the bait for a period of time before taking it, so you should wait a few beats before setting the hook.

Table Fare: Variable. Poor for large drum that have coarse and tough flesh, which may also contain parasites (spaghetti worms) that are not harmful to humans but rather unappealing. Good for drum less than about ten pounds.

DRUM, RED (*SCIAENOPS OCELLATUS*), CHANNEL BASS, REDFISH

N.C. Citation Size: Release only, forty inches or longer.

Availability: April–May and October–November (large drum), April–December (puppy drum).

Coloration: Red drum derive their name from their distinctive coloration that ranges from a deep copper to almost silver. Most red drum have a reddish-brown back that fades to a white belly.

Distinctive Features: Red drum have a spot on either side of the base of the tail, but some fish may have two or more spots. The species name *ocellatus* is Latin for "eyelike colored spot." Adults have a long body that tapers in the front to a blunt head and in the back to a truncated caudal fin. The mouth is underslung—slightly underneath and behind the snout.

Diet: A drum's underslung mouth is a major factor in its food preference: bottom critters, such as shrimp and crab, which drum efficiently root out from the soft, sandy bottom. But the red drum is not averse to chasing down baitfish, such as finger mullet that migrate down the coast during the fall.

Biology & Behavior: Red drum are members of the family Sciaenidae that includes their seatrout, croaker, and black drum cousins. During spawning time, males produce a drumlike sound by vibrating a muscle in their swim bladder. These fish are prolific spawners; large females produce nearly 2 million eggs in a single season. Spawning occurs in the waters around coastal inlets and in some areas of Pamlico Sound during late summer and fall.

For the first three to four years of their life, red drum reside in estuarine waters bordering the Outer Banks. As red drum mature into juveniles, commonly referred to as puppy drum (eighteen to twenty-five inches in length), they are found scavenging along the ocean side of the coast. Large, mature drum, which may live to be sixty years old, prefer deeper waters and tidal inlets, whereas smaller drum remain in shallower water.

Probably more big red drum are found along the Outer Banks than along any stretch of the Atlantic coast because of the rich and abundant feeding opportunities. These giants are not highly migratory, preferring to move offshore as coastal water temperatures drop during the fall and return in the spring.

Angling Techniques: Since drum routinely feed on the bottom, particularly in rough water, where they find most of their meals by scent and feel, bottom-fishing is de rigueur for catching trophy red drum in the surf or from a pier. Fresh mullet or menhaden are impaled on large (6/0 to 9/0) circle hooks that are used by the locals who regularly target red drum. Stout rods (ten to twelve feet in length), reels with twenty-pound test line, and a shock leader of forty- to sixty-pound test allow anglers to handle these hard-fighting fish.

Angling Tip: Many anglers fish after dark for large drum, which are mainly nocturnal feeders. A southwesterly wind and an incoming tide appear to be ideal conditions that often stimulate large schools of drum to move shoreward within casting distance of wading anglers. (When fishing Cape Point at night, dim your vehicle's lights and never shine your lights on the water; otherwise, you may shut down the bite.)

Table Fare: Good. Small red drum, less than ten pounds, have a sweet, mild flavor.

Flounder, Summer

FLOUNDER: SUMMER, SOUTHERN, AND GULF (*PARALICTHYS DENTATUS, P. LETHOSTIGNA, P. ALBIGUTTA*)*

N.C. Citation Size: Five pounds.

Availability: May–October.

Coloration: Body coloration and patterning change with the background, but the flounder is typically brownish with a variable number of spots. Its belly or underside is white.

Distinctive Features: When viewed from above, a flounder has both eyes on one side of its body. In particular, all three flounders are left-eyed fish. All flounder have a depressed (flattened) body shape that is ideally suited for dwelling on the bottom.

Diet: With their well-developed teeth, flounder feed on small fish, squid, seaworms, shrimp, and other crustaceans.

Biology & Behavior: Flounder are the epitome of ambush predators: partially burrowing into the sediment, patiently waiting for their prey, and focusing their eyes upward. With a thrust of their broad tail, they can quickly spring out from their concealment to overtake an unsuspecting food item. Most flounder visit the coastal waters of North Carolina from spring to autumn and then migrate

*Though there are three distinct species that inhabit North Carolina waters, they have similar anatomical, morphological, and behavioral traits. I will treat them as one.

offshore during the winter months to spawn. Spawning begins at age two when the fish are approximately twelve inches in length and generally occurs during the winter migration. Most flounder species can live at least two decades, but females live longer and grow bigger than males.

Angling Techniques: Whatever your choice of lure or bait, your success depends mainly on moving your offering through the target zone. Though fishing stationary bait can catch flounder, remember that flounder are very adept at ambushing their prey; they expect a fleeing victim. Move the rod in a slow, steady sweeping motion, pausing only briefly before beginning again. While this technique is particularly effective in giving "life" to a long strip of bait, such as squid, leadhead jigs with a plastic tail can be bounced along the bottom by vertically twitching the rod tip. A seven-foot rod, which is matched with the appropriate spinning or bait-casting reel (six- to ten-pound test), is sufficient to catch these tenacious fish. Many anglers use a flounder rig—a wide-gapped hook adorned with a spinner blade and a feathered dressing—and fish a live minnow or finger mullet. From the beach, long casts are not necessary because most flounder will be lurking in near-shore drop-offs and waiting for their next meal to tumble down to them.

Angling Tip: A short piece of leader material inserted between the line and the hook is helpful in negating the flounder's two rows of sharp teeth.

Table Fare: Excellent. Flounder may be prepared in many ways: broiling, frying, or baking. Some gourmands prefer their flounder stuffed with crabmeat.

JACK, CREVALLE (*CARANX HIPPOS*)

N.C. Citation Size: Release only, thirty-two inches or longer.
Availability: June–September.
Coloration: The fish's back is greenish-blue or bluish-black, and its
    underside is silvery white to yellowish. There is a distinct black
    spot on each gill cover.
Distinctive Features: The crevalle jack has a large, rounded head with
    saucerlike eyes. The body is generally smooth except for a small
    patch of scales in front of the pelvic fins, and the body tapers to a
    pronounced forked caudal fin.
Diet: Crevalle jacks feed during the day upon a variety of small fish
    and invertebrates.
Biology & Behavior: The spawning season of crevalle jacks is from
    March to September. Spawning sites include offshore locations
    in the southeastern Atlantic and the Gulf Stream. When young,
    crevalle jacks congregate in large schools, but with maturity, they
    become solitary in nature and are most often found offshore in
    depths of up to 100 feet.

    Crevalle jacks have been known to ingest a toxin that is pro-
    duced by the phytoplankton *Gambierdiscus toxicus*. Though the
    toxin is not harmful to the fish, human consumption of the flesh
    of crevalle jacks can possibly result in ciguatera poisoning.

As voracious predators, crevalle jacks have been observed to corral schools of baitfish at the surface, or they will pursue their quarry well into the shallow water.

Angling Techniques: There are two basic ways to catch big crevalle jacks. As with king mackerel, live-bait fishing is the preferred method of pier anglers. Hooking any small fish through the back, in front of the dorsal fin, will prove to be successful if a jack is in the vicinity of the offering. Casting lures can be another effective approach. Of the many game fish that swim past the Outer Banks piers, jacks will strike at almost any type of lure, such as top-water plugs, jigs, plastic grubs, and metal spoons. (On crowded piers, this method is not advised because of the likelihood of crossing lines and causing entanglements.)

Angling Tip: When working a lure, a high-speed retrieve is usually most productive. Jacks seem to love the pursuit, and a slow-moving lure offers little appeal to these aggressive predators. If the jack misses the lure on the first pass, keep moving the lure to keep the jack's interest.

Table Fare: Poor. Jacks are generally held in low esteem as a food fish and thus released.

Kingfish, Southern

KINGFISH: NORTHERN, SOUTHERN, AND GULF (*MENTICIRRHUS SAXATILIS, M. AMERICANUS, M. LITTORALIS*); SEA MULLET, WHITING*

N.C. Citation Size: One and a half pounds.
Availability: April–May, October–November.
Coloration: Body is silvery gray or tan dorsally, and belly is silvery white. A series of wide and faint (brownish) bars are found on the side of its body. On the northern kingfish, these bars are more distinct.
Distinctive Features: Chin with single, short barbel; elongated and rounded body, terminating with a truncated caudal fin; nine to ten spines on first dorsal fin.
Diet: Kingfish are essentially bottom-feeders, primarily consuming crustaceans and polychaete (segmented body with bushylike protrusions) worms.
Biology & Behavior: Kingfish are demersal, occurring over a wide variety of bottoms from mud to sand-mud mixtures; however, adults frequent the sand bottoms of ocean beaches, and juveniles inhabit the surf zone and estuarine waters. Their underslung mouth and pointed snout are ideal for rooting around the bottom in search

*North Carolina is home to several very similar species of kingfish, including northern, southern, and gulf kingfish.

of burrowing organisms. They are found in a wide range of water temperatures (45°–85°F).

Spawning occurs between ages two and three from May to August in coastal waters. The species is known to live at least four years.

Angling Techniques: Leave the lures at home; these inshore species are an easy target for a surf caster or pier angler who is armed with a standard two-hook bottom rig. Hook size (#2 to #4) is critical because these fish have small mouths. If you are detecting strikes but not hooking up, then simply switch to smaller hooks. Bloodworms and shrimp top the bait menu, but a fresh piece of squid or a mole crab catches its fair share of these small but hard-fighting fish.

Deep sloughs bordered by sandbars appear to hold most of the fish. Because these fish-holding features are generally no more than twenty to thirty yards out from the beach, medium spinning outfits with eight- to twelve-pound test line are sufficient. As in most angling situations, sinker weight depends on wave activity and strength of current, but generally a two- to four-ounce pyramid sinker is sufficient to hold bottom.

While sea mullet don't congregate in huge schools, when they are feeding actively, the action can be almost nonstop, requiring the angler to hold the rod instead of using a sand spike.

Angling Tip: While water clarity may be a key to catching some surf species, sea mullet will bite in both clear and dirty water. Also, I have known some anglers to have quite a bit of success fishing for sea mullet at night.

Table Fare: Good. White, flaky meat is very firm and tasty.

MACKEREL, KING (*SCOMBEROMORUS CAVALLA*)

N.C. Citation Size: Thirty pounds or live release of fish thirty-five inches or longer.

Availability: May-October.

Coloration: The back has iridescent tones of blue and green that fade to silvery white on the belly. Most of the fins lack any distinct coloration, except the first dorsal fin, which is uniformly blue.

Distinctive Features: A highly streamlined body that tapers to a pronounced forked caudal fin. A distinctive lateral line starts high on the back and drops sharply below the second dorsal fin. The king's large, cutting-edged teeth, of which there are thirty, are closely spaced and flattened from side to side.

Diet: Like other members of the genus, including Spanish mackerel and cero mackerel, king mackerel are carnivores, preferring to slash through and feed on schools of fish, such as grunts, snappers, and jacks.

Biology & Behavior: King mackerel can be found cruising coastal as well as open waters, preferring temperatures that seldom drop below 67°F. But they may also be found around piers, buoys, wrecks, and other sites when prey is abundant. During the year, king mackerel migrate extensively along the eastern seaboard in schools of various sizes, although the larger specimens are generally solitary.

At least two stocks of king mackerel are known to scientists: one found in the Gulf of Mexico and the other in the western Atlantic. The Atlantic stock is particularly abundant off North Carolina in the spring and fall but migrates south to spawn from May through August, slowly returning to the mid-Atlantic region throughout the summer. Evidence points to this stock wintering in deep water off the Carolinas.

Angling Techniques: Known for its blistering run when hooked, the king mackerel is one of the most coveted species that inhabits the waters of the Outer Banks. They are mostly taken from boats trolling various live and dead baits, spoons, and lures. For many Outer Banks charter captains, the preferred method is to live-bait menhaden on treble hooks with twenty- pound tackle.

Pier anglers also catch their fair share of kings by soaking live bait at the end of the pier. Baits of choice include spot, pinfish, menhaden, and small bluefish, but in a pinch, almost any live bait can be and has been used by king mackerel fishermen. Typically with live bait, two hooks are tied to a strong metal leader. (Many tackle shops carry pre-made kingfish rigs.) The first hook may be a treble or single and is positioned before the first dorsal fin. The second hook (treble) is placed through the top of the fish's back near its tail or allowed to swing free. This setup is employed because king mackerel are notorious for their short strikes—cutting the bait in half just behind where the first hook was inserted.

Angling Tip: Since king mackerel are vision-dependent predators, use small hooks and light leaders to minimize the rig's appearance. While this setup will generally entice the larger but more wary king to strike, the downside is that tiny treble hooks may often pull on the king's initial run.

Table Fare: Good. While small fish are generally grilled, specimens larger than twenty-five pounds are best smoked. Big king mackerel have high iodine and oil content that imparts a strong flavor to the flesh, so many anglers will have the fish smoked.

MACKEREL, SPANISH (*SCOMBEROMORUS MACULATUS*)

N.C. Citation Size: Six pounds.

Availability: June–September.

Coloration: They are green or greenish-blue on top, fading to silver sides and belly; a large number of irregular golden-olive spots adorn the sides. Juvenile king mackerel can be mistaken for adult Spanish mackerel because they also have these distinctive body spots, which are not present in adult kings. A distinguishing characteristic is the dark black front portion of the Spanish mackerel's first dorsal fin.

Distinctive Features: A Spanish mackerel's body is elongated and strongly compressed from side to side; the body is nearly five times as long as it is deep. However, the most prominent anatomical feature of a Spanish mackerel is the large, conical jaw teeth. The caudal fin of the Spanish mackerel is markedly lunate, and its caudal peduncle is keeled. The pronounced lateral line curves smoothly downward to the base of the tail.

Diet: Spanish mackerel are schooling fish that relentlessly pursue and attack small herringlike fish, including bay anchovies, glass minnows, and silversides.

Biology & Behavior: The Spanish mackerel is built (streamlined body and lunate caudal fin) for speed, traveling over thirty miles per hour in short bursts. Feeding Spanish mackerel will often leap clear of the water in pursuit of fleeing prey.

    Spanish mackerel are highly migratory, moving hundreds of miles from their wintering grounds in southern Florida to as far north as New York during the summer. During this migration, they form large, fast-moving schools, preferring to stay in waters with temperatures above 68°F.

Spanish mackerel spawn over a protracted season—April in the Carolinas and late September off Sandy Hook, New Jersey. Few mackerel spawn when sea surface temperatures drop below 75°F. Though there are no reliable measurements of growth rates in Spanish mackerel, particularly juvenile fish, researchers have estimated their life span to be five to eight years.

Angling Techniques: The Spanish mackerel is a summertime favorite of surf, pier, and inshore anglers, who all use a variety of lures to entice strikes from these speedy predators. Surf anglers cast small (one to three ounces), flashy, aerodynamically shaped lures, such as Stingsilvers. Nine-foot spinning rods and spinning reels, spooled with ten-pound test line, allow anglers to cast their lures to distant fish. The retrieve needs to be fast to imitate a fleeing prey. (You can't reel fast enough for a determined mackerel.) Got-cha plugs are the choice of pier anglers, and boaters troll silver or gold Clarkspoon lures, positioned down in the water column with planers.

Being sight feeders, Spanish mackerel shy away from turbid, roiled water. Confine your efforts to clear water. While random casting will often catch a cruising mackerel, target schools of fish that are feeding under a flock of birds that are diving down to pick up the morsels. For the shore-bound angler, early morning and evening are generally the most productive times, as these fish move in from the deeper waters to forage in the shallows during low-light conditions.

Angling Tip: Since Spanish mackerel have excellent vision, avoid any terminal tackle (snaps) between your line and lure. If the mackerel are particularly wary, a two-foot length of fluorocarbon leader (essentially invisible in the water) attached to the lure may tip the scales in your favor.

Table Fare: Good. Fillets are best either grilled or broiled.

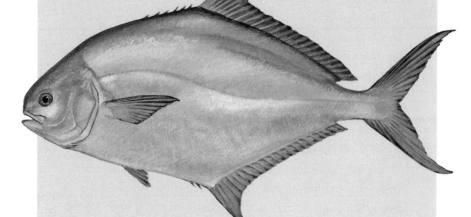

POMPANO, FLORIDA (*TRACHINOTUS CAROLINUS*)

N.C. Citation Size: Two pounds.

Availability: June–September.

Coloration: Greenish-gray on back, shading to silvery sides, and with a bright yellow underbelly.

Distinctive Features: A deep, compressed body tapers to a blunt head. The mouth of a pompano is small and is devoid of any teeth. The caudal fin is broadly forked, and the second dorsal fin has twenty-two to twenty-seven soft, prominent rays.

Diet: Pompano feed on shrimp, crab, clams, mussels, and sand fleas (mole crabs).

Biology & Behavior: Schools of pompano, ranging in size from small to large, migrate north in summer along the Atlantic seaboard. They are usually found in the clear, shallow waters along sandy beaches and prefer water temperatures in the mid-80s. Anglers often see pompano in waters no more than a few inches deep, as these fish root in the sand for mollusks and crustaceans, which they crush with their strong jaws.

Spawning occurs offshore in late spring to summer, and the juveniles move inshore to low-wave-energy beaches. Most of the pompano caught during the height of summer are these young fish, with the bigger (two to three pounds) and older fish (two to four years) making an appearance early in the spring and again in late summer.

Angling Techniques: While pompano are caught by pier anglers, most are taken by fishermen prowling the beach searching for signs of pompano activity. In particular, anglers look for a concentration of sand fleas—pompanos' favorite food—which have been exposed by the receding wave. The scurrying sand fleas can easily be collected by means of a sand flea rake (available at local tackle shops) and kept for hours in a bucket filled with wet sand.

Pompano travel in schools, and if you catch one, you're likely to catch more. Constantly on the move in search of food, pompano never remain in one location for long. Their movement appears to be controlled by both current and tide. With regard to the latter, an incoming tide is the best time to target pompano, as they surge over the outer bar with the incoming water to forage in the nearshore trough. If the water is clear—a preference of pompano—a school can be seen in the break of the wave.

Most surf anglers employ two rods: one long rod (ten-foot) to fish the back side of the outer sandbar and a shorter rod (seven-foot) to fish the swash zone where waves break on the beach. If you are fishing from a pier, avoid the end of the pier and concentrate on the nearshore surf break.

A two-hook (size the hook to the pompano's small mouth) bottom rig is standard, and some anglers often add orange beads that, in theory, simulate a sand flea laden with a mass of orange-colored eggs. The sand flea is generally hooked from the rear. Use an appropriately sized pyramid sinker to hold bottom.

Angling Tip: Keep the hooks sharp. Even the most expensive hooks will quickly dull due to the combined effects of being pushed repeatedly through a sand flea's hard shell and tumbling and dragging in the surf.

Table Fare: Excellent. Firm, white flesh, particularly when baked, is a favorite of many anglers.

SEATROUT, SPOTTED (*CYNOSCION NEBULOSUS*), SPECKLED SEATROUT, SPECK

N.C. Citation Size: Five pounds or live release of fish twenty-five inches or longer.

Availability: November–December (surf), April–November (sounds).

Coloration: The dorsal side of the body is silvery and has irregular black spots from the first dorsal fin to the caudal fin. The ventral side is silvery white and lacks any spots.

Distinctive Features: The spotted seatrout has an elongated, somewhat compressed body. The head is long and has a pointed snout and large oblique mouth containing canine teeth. The dorsal fin is continuous or slightly separated along the trout's back. The interior of its mouth is bright orange.

Diet: Though newly hatched trout feed mainly on small zooplankton, there is a dietary shift with age to larger items, such as mysids and shrimp. Large trout feed on small fish (silversides, menhaden) and crustaceans.

Biology & Behavior: The species is a member of the Sciaenidae family of croaker and drum, and its genus name, *Cynoscion*, refers to the tender membrane in its mouth. Spotted seatrout are very fecund; large females may produce 15,000 to 1 million eggs per nocturnal spawning event. During this period, the males signal their interest by emitting croaking sounds one to two hours prior to sunset. Researchers have determined that environmental factors, such as water temperature and salinity levels, are more important than specific location in spawning success.

Juveniles spend their first couple of years in the marshy environs of the sound and then move into the inlets and along the beaches. Adult trout travel in small schools with the incoming tide and feed in the shallow sloughs. They are ambush predators and make short lunges to grab prey with their front canine teeth.

Though spotted seatrout can tolerate a relatively wide range of temperatures (58° to 81°F), cold water is lethal to them. While some may seek the comfort of deeper water, trout that remain in the shallows of the sound may be killed during abnormally cold winters.

Angling Techniques: Spotted seatrout are schooling fish, and once they are located, they are relatively easy to catch. These fish prefer holes or depressions where they can corral baitfish, which do not have much chance of escape. Surf anglers frequently find trout close to the beach, lurking in deep sloughs. Keep in mind that the nearshore environment is constantly changing, and trout-holding sites may migrate along the beach in response to tides and currents. Take time to search out the beach. Sound anglers also target deep holes that pockmark the bottom of the relatively shallow sounds of the Outer Banks. In warmer weather, these deeper holes, in addition to holding bait, will have more oxygen and be cooler than the surrounding shallower sections. During the spring and autumn, spotted seatrout can be taken at high tide over eelgrass beds by using either bait (peeler crab, live shrimp) or lures.

A seven-foot spinning rod (six- to eight-pound test line) is the choice for those who target trout. Anglers cast three-eighths to half-ounce leadhead jigs, which have plastic, curly-tail grubs on their hooks, into the nearest slough and slowly retrieve them. (Trout aficionados constantly debate the choice of jig color, but a red head and a green tail seem to be the most popular.) The key to success is to impart a slow, hopping motion to the lure that triggers the predatory instincts of the trout. Twitching the rod tip upon retrieve allows the lure to move erratically through the water column. Keep alert; trout often strike the lure on its descent. MirrOlures, plastic imitations of small baitfish, have also caught their share of these fish. Many anglers use the standard M-series MirrOlure, which sinks slowly, and they retrieve it with a slow, twitching sweep of the rod.

Angling Tip: Though speckled trout can be caught throughout the day, low-light conditions, sunrise and sunset, appear to be most productive for surf anglers. Couple this time of day with an out-going tide, which concentrates prey and predator alike in the deep holes, and you have the combination for success.

Table Fare: Good. The fish has a delicate flavor that will benefit from the simplest methods of cooking.

SHARK, BLACKTIP (*CARCHARHINUS LIMBATUS*)*

N.C. Citation Size: 150 pounds or live release of any shark eighty inches or longer.

Availability: June–November.

Coloration: The blacktip shark is dark gray to brown above and white below, with a distinctive white stripe on each flank. As the name implies, the pectoral fins, the second dorsal fin, and the lower lobe of the caudal fin usually have black tips, though they tend to fade with age.

Distinctive Features: This shark species has a stout, streamlined body with a long, almost V-shaped snout. The first dorsal fin is high and tapers to a narrowly pointed apex. In contrast, the second dorsal fin is small, and there is no ridge between the two dorsal fins. As with most sharks, blacktips have broad, winglike pectoral fins behind the five gill slits. The upper and lower jaw teeth of these sharks are similar in appearance—broad-based, moderately long, and coarsely serrated.

*According to Frank Schwartz, a noted shark expert, there are more than two dozen shark species that are common to the waters of the North Carolina coast. These include pelagic species, such as the shortfin mako shark, which roam the Gulf Stream, and inshore species, such as the sand tiger, sandbar, and spinner sharks. Even the great hammerhead has been known to venture inshore during the summer months. In 1960, a 710-pound individual was caught from the Nags Head Fishing Pier. In addition, curators at the North Carolina Aquarium point out that the coastal waters are the birthing grounds for six kinds of sharks.

Diet: These ravenous feeders consume a wide variety of small fish, including herring, anchovies, mullet, and menhaden. They also feed on skates, rays, and smaller sharks, such as sharpnose and young dusky sharks.

Biology & Behavior: The blacktip shark belongs to the family of requiem sharks that includes the tiger, blue, and bull sharks. The name "requiem" comes from the French word for shark, *requin*. They prefer inshore habitats and are found in water less than 100 feet deep. Although an individual may be found a considerable distance offshore, this shark is not a truly pelagic species. Their seasonal migration off the east coast of the United States includes moving north to North Carolina in the summer and south to Florida in the winter. They generally travel in groups of varying size and sex; adult males and nonpregnant females are found apart from pregnant females, and both are separated from juveniles.

The blacktip shark is an extremely fast, acrobatic predator. Like the spinner shark, this species is known to leap out of the water while feeding, often spinning three or four times around its axis. Some of these jumps are the result of the shark's feeding strategy: it corkscrews vertically through a fish school at high speed, and its momentum carries it through the ocean surface.

As with other requiem sharks, reproduction in the blacktip is viviparous: it gives birth to live, free-swimming young. Females typically give birth to four to seven pups each year in shallow coastal estuaries, where the young remain for the first years of their lives. At maturity, blacktips attain an average length of five feet and a weight of approximately fifty pounds, but the all-tackle world record is a 270-pound, 9-ounce fish taken off Kenya.

Angling Techniques: While shark fishing from the piers of the Outer Banks is now generally discouraged because of the inherent dangers, sharking from the beach has essentially taken its place. Baits such as mullet, menhaden, and small tuna carcasses are carried out hundreds of yards to the deeper water by means of a kayak or similar vessel. While heavy tackle is not needed for blacktips, you should have a variety of rods and reels in place in case a large shark shows up. (Sand tiger sharks, which periodically cruise the surf zone, range in size from six to nine feet and reach weights of over 300 pounds.) While large spinning reels are adequate for smaller species, conventional reels in the 6/0 to 12/0 range allow the angler to pack on more line and to handle the bigger brutes. All reels should be capable of holding hundreds of yards of thirty- to fifty-pound test line and are matched with medium- to heavy-action rods in the five- to seven-foot range. A twenty- to thirty-foot stainless steel leader terminates to a large hook (12/0). (For large baits, some anglers employ a two-hook rig.) The last piece of gear should be a fighting belt. Blacktip fishing, as with most shark fishing, requires patience, but once hooked, the species is a strong, steady fighter, often leaping out of the water.

Angling Tip: Fishing for these beasts is not a solitary venture; teamwork is required to land and ultimately release these potentially dangerous sharks, with safety being the first priority for all participants. In that vein, anglers often use cable cutters to cut the hook from the leader rather than attempting to remove it.

Table Fare: Good. Sharks retain urea in their bodies, so if not properly cared for (eviscerated and iced), they will spoil very quickly after death.

SHEEPSHEAD (*ARCHOSARGUS PROBATOCEPHALUS*)

N.C. Citation Size: Three pounds.

Availability: May–October.

Coloration: Basic silvery color, with five or six distinct vertical black bands along each side. Because of this banding and their propensity to "steal" bait off the hook, sheepshead have earned the nickname "convict fish."

Distinctive Features: Prominent teeth, including eight broad, incisor-like teeth lining the front of both jaws and three rows of molars in the upper jaw and two rows in the lower jaw. The sheepshead derives its name from its large, sheeplike incisors. Rigid and sharp spines are found on the dorsal and anal fins.

Diet: Sheepshead feed on a variety of crustaceans, including crabs and barnacles, and mollusks, such as clams. A sheepshead's formidable dental work can scrape barnacles from pilings and crush the hard shells of crabs and mollusks.

Biology & Behavior: Sheepshead are notoriously structure oriented; they hold close to the pilings of the pier, where they remain in the shade of the structure and dine on the attached organisms. Though little is known regarding spawning behavior, the limited data show adults migrating to offshore waters in the early spring to spawn, later returning to the nearshore waters. The maximum known life span of the sheepshead is at least twenty years.

Angling Techniques: Choosing and attaining the appropriate bait is the first step in fishing for sheepshead. Most seasoned anglers prefer fiddler crabs, which can be found scurrying around the shoreline of marshy creeks during low tide. In the absence of fiddler crabs, try sand fleas (mole crabs), which burrow into the wet sand left by receding waves. Upon procuring the bait, use a stout hook, matched to the size of the bait (generally 1/o to 2/o hooks), because sheepshead can bite through thin wire hooks. Since success depends on getting the bait as close to the piling as possible, tie the hook directly to a loop on the line, attach a sinker a foot below the hook, and drop the rig straight down to the bottom. Then, reel the rig in a couple of feet and proceed to slowly jig your offering. By moving the bait up and down, you decrease the chance of the sheepshead sneaking in and stealing your bait, because this controlled, deliberate movement allows you to detect any change in the bait and set the hook immediately with a quick snap of the rod. Tighten the drag and be prepared to take control of the fish when it bulldogs its way to the structure.

Angling Tip: Since the sheepshead's bite can be very subtle, some anglers have switched from monofilament to braided line because of its increased sensitivity.

Table Fare: Excellent. Novice anglers may find these fish difficult and time-consuming to clean.

SPADEFISH (*CHAETODIPTERUS FABER*)

N.C. Citation Size: None at present.

Availability: April–June.

Coloration: The fish is silvery and has four to six black vertical bands on each side of its body. The first band runs through the eye, and the last band is near the tail. In larger fish, these bands may fade or become obscure. Though adult coloration is similar to that of sheepshead, spadefish are morphologically different, and the black bars do not fade in sheepshead.

Distinctive Features: The spadefish is a laterally compressed, disk-shaped fish with two distinctly separated dorsal fins. The first dorsal fin is small and spiny, and the second fin, which is soft, is large and elongated to the rear of the fish. The head tapers to a blunt snout and a small mouth.

Diet: Spadefish are extremely generalized in their choice of food, consuming a wide variety of benthic organisms (crustaceans,

mollusks, and worms), as well as organisms (jellyfish) throughout
the water column.

Biology & Behavior: Researchers have determined that spadefish reach
a maximum length of almost three feet and a peak weight of approximately eighteen pounds. (The pending International Game
Fish Association all-tackle world record is over fourteen pounds.)
The expected lifespan is eight to ten years. Sexual maturity for
both sexes is approximately one year of age. Spawning occurs on
the inner continental shelf from May through September, with
a single female capable of releasing up to 1 million eggs. Anglers
may come upon large spawning aggregations near the surface,
particularly on warm, sunny days when water temperature is 75°
to 85°F.

Juvenile spadefish can be found in shallow water, often cluster-
ing around pier pilings. In contrast, adults may school into groups
of up to 100, seeking out depths of fifteen to twenty-five feet. This
species can be found clustered around various navigation markers
or anywhere there is structure.

Angling Techniques: More and more anglers appear to be targeting
spadefish in the coastal waters of North Carolina, and with good
reason: spadefish are strong fighters. Because these fish are structure oriented, accurate positioning of your boat in relation to the
structure is critical to success. Since the weather on the Outer
Banks can be quite changeable, choose a day when seas are calm
and current flow is minimal. These conditions will allow you to
anchor, which is preferred over drifting.

Clam is the bait most anglers prefer, using big, fresh pieces
when large spadefish are around. Hook the bait through the tough
foot section of the clam and let the rest dangle free. Since a natural
prey of spadefish is jellyfish, some anglers employ it when targeting spadefish, but catching and handling jellyfish can be tricky as
well as messy.

Match the hook size to the size of the spadefish—#6 hook
for small fish and up to #1 for the larger specimens. Tie the hook
directly to a two-foot section of twenty-pound test leader. On
the other end of the leader, a swivel forms the link between the
leader and the main line. Before tying on the swivel, thread an egg
sinker onto the leader. The weight of the sinker is determined by

the depth that the fish are holding, but generally an ounce or two will do the trick. Lower the baited rig to the depth of the fish. You may have to experiment a bit to determine the appropriate depth, but ten to twenty feet is a good starting point. Put the rod in the rod holder, with the reel set on a relatively light drag, and let the fish hook itself. If you decide to hold the rod, don't set the hook upon detecting the bite; pulling too hard can tear the hook from the spadefish's relatively delicate mouth. As long as you can detect the fish nibbling at the bait, there is a good chance that it will soon hook itself. If the bite stops, the spadefish has made off with your bait.

Angling Tip: Chumming with frozen clams can bring the fish closer to your boat and may turn on the group that until now has spurned your offerings.

Table Fare: Good. The flesh is predominantly white and mild tasting.

SPOT (*LEIOSTOMUS XANTHURUS*)

N.C. Citation Size: One pound.

Availability: August–October.

Coloration: Light bronze on the upper back, shading to white on the underbelly. They also have twelve to fifteen dusky oblique bars on their upper side.

Distinctive Features: The spot's small body is stout and deep, and the tail is slightly forked. The dorsal fin is soft and slightly notched. A relatively large, black spot—thus its name—is located above each pectoral fin.

Diet: Like croaker, spot are opportunistic bottom-feeders, consuming worms, small crustaceans, and mollusks.

Biology & Behavior: Being the smallest members of the Sciaenidae family, spot play key roles in the food chain dynamics of North Carolina's coastal waters: predators of bottom invertebrates and as a major prey species of striped bass, bluefish, and flounder. Although both spot and croaker share the same habitat and have a similar diet, they are able to coexist without directly competing for resources.

Spot reach sexual maturity between ages two and three, but they are relatively short-lived and rarely reach the five-year mark.

Spawning occurs in offshore waters during the late fall, when water temperatures start to drop. Female spot are very fecund, capable of producing over 1 million eggs. The eggs are carried shoreward by winds and currents, and the resulting juveniles are estuarine dependent. Upon reaching maturity, spot return to the ocean.

Angling Techniques: A standard two-hook bottom rig, tipped with bloodworms—generally spots' preferred meal—is used by many experienced pier anglers. Spot often travel in a large school (over 100), so it is not unusual for the action to be fast and furious—two fish at a time—and before long you can have a cooler full of these tasty bantamweights of the seashore. While many pier aficionados are content to drop their rig over the railing and let it sit on the bottom, gently lifting and lowering the rod, which imparts some "action" to the offering, often triggers a strike.

Angling Tip: If you are getting bites but are not hooking fish, try downsizing your hooks, since spot have extremely small mouths.

Table Fare: Excellent. What these fish lack in size they make up for in taste, particularly when pan fried.

TARPON (*MEGALOPS ATLANTICUS*)

N.C. Citation Size: Live release of fish, regardless of size.

Availability: July–August.

Coloration: Predominantly bright silver along its sides and belly, prompting many anglers to refer to the tarpon as the "silver king."

Distinctive Features: The tarpon is almost prehistoric in appearance; its huge mouth, in particular, has a projecting, upturned lower jaw that contains an elongated bony plate. The tarpon's silvery sides are made up of large, platelike scales that appear to flash in the sunlight when the fish jumps into the air. Perhaps the most unique internal feature of the tarpon is a modified swim bladder that allows the fish to extract oxygen directly from the atmosphere and increases its chance of survival in oxygen-poor waters.

Diet: While juvenile tarpon mainly consume tiny zooplankton, such as copepods, adults almost exclusively feed on fish, such as mullet and pinfish, and benthic organisms like shrimp and crab.

Biology & Behavior: Although a marine species, tarpon is a euryhaline organism: it is able to tolerate large changes in salinity. Their habitats can stretch from the ocean to brackish bays to freshwater creeks. The only variable that limits their habitat and range is temperature, with rapid decreases in temperature resulting in large fish kills. Recent research has shown tarpon are thermophilic, requiring high temperatures for normal development. Tarpon are slow-growing fish, only reaching sexual maturity by the age of six or seven and a length of four feet. Male tarpons have a life span of over thirty years, and females may live longer than fifty years. A female in a Chicago aquarium died at age sixty-three.

The eggs released by a spawning tarpon develop into larvae called leptocephali—strange-looking creatures that have transparent, snakelike bodies and fanglike teeth. Upon drifting into estuarine areas, the larvae lose their teeth and shrink in size, evolving into miniature versions of their adult counterparts.

Angling Techniques: The primary tarpon destination is Pamlico Sound, particularly the western portion around the mouth of the Neuse River. (That said, tarpon are routinely caught from piers, but the anglers were not specifically targeting these fish.) As opposed to the clear waters of the Florida Keys or the Caribbean that allow the angler to sight cast to a cruising tarpon, fishing for tarpon in the murkier Pamlico Sound involves arriving at a predetermined spot, anchoring, setting out baits, and waiting for a strike. Whole dead spot, croakers, menhaden, or bluefish are fished on the bottom, and many successful anglers choose stout boat rods (five to six feet in length) and thirty- to fifty-pound test line to quickly land the fish. Though circle hooks are a good choice to prevent gut-hooking the fish, by all means make sure they are sharp, because they must penetrate the hard, bony jaw of the tarpon. Generally, only one strike in ten results in a hookup, and of those hookups the tarpon throws the hook at least half the time.

Angling Tip: "Bow" to the silver king. During the fight, many tarpon break the line when they fall back into the water after a spectacular leap. If too much tension is kept on the line during the leap, the line easily parts under the weight of the fish. By bowing, the angler gives enough slack to the line.

Table Fare: Poor.

TUNA, BLUEFIN (*THUNNUS THYNNUS*)

N.C. Citation Size: Eighty pounds or release of fish, regardless of size.

Availability: December–February.

Coloration: The body is metallic deep blue on top, and the lower sides and belly are silvery white.

Distinctive Features: This species is the largest of the tunas; giant bluefin tuna reach a minimum length of eighty-one inches and a weight of 310 pounds. Bluefin have a fusiform body; it tapers at both ends from a relatively thicker midsection. Compared to other tuna, its head is long and somewhat pointed, with small eyes. There are two distinct dorsal fins, with the second taller than the first. Behind the second dorsal fin and the anal fin are seven to ten finlets. The caudal fin is lunate in shape.

Diet: Bluefin are eclectic feeders that target small schooling fish, particularly anchovies and herring, as well as starfish, kelp, and smaller fish in shallow water. But in North Carolina during winter, medium and giant bluefin tuna will feed almost exclusively on adult Atlantic menhaden.

Biology & Behavior: The only confirmed spawning ground for bluefin tuna in the western Atlantic is the Gulf of Mexico. The females shed their eggs in the spring, and the 2010 oil spill in the Gulf of Mexico occurred simultaneously with this spawning event. What will be the long-term effect on this species? Tagging studies have shown that bluefin tuna are highly migratory and ride the Gulf Stream annually to their summer feeding grounds in New England and Canadian waters. During these journeys, bluefin will often travel in a parabolic (arc) formation that allows them to wrap around a ball of baitfish. Bluefin are uniquely adapted for these long journeys: they are endothermic (maintain an elevated body temperature), have large gill surface areas to extract dissolved oxygen from the water, have large hearts to pump the oxygenated blood to their tissues and muscles, retain 98 percent of their muscular heat, and have a highly streamlined body to minimize water drag.

Bluefin tuna are especially vulnerable to overexploitation because they are slow to mature and are an economically valuable resource. Bluefins in the western Atlantic reach sexual maturity at approximately age eight (eighty inches in length). In 2009, the total bluefin commercial catch in North Carolina was 138,902 pounds and sold for almost $1.5 million.

Angling Techniques: In the mid-1990s, commercial fishermen discovered large concentrations of bluefin tuna over wrecks about twenty miles from Hatteras Inlet. Since then, anglers have targeted these wrecks, using cut bait such as menhaden to chum the fish up to the surface. They simply hook up a chunk of bait and toss it over. But over the years, increasing fishing pressure has led to more wary fish, so it is a good idea to use lighter leaders and conceal the hook from sharp-eyed tuna. Recent surveys have shown that the number of fish returning to waters off Hatteras has been decreasing, and anglers are resorting to trolling two to four lines with eighty-pound tackle. If a fish is hooked, and a school is located, anglers change to chunking—tossing over the chunks of cut bait.

Angling Tip: Many charter captains use the heaviest tackle possible, 130-pound outfits, because quite often the fish top out over the size limit that can be kept. The heavier tackle allows the angler to exert more pressure on the fish and ensures a quick release that increases the tuna's chance of survival.

Table Fare: Fair when cooked, but if your tastes gravitate toward raw fish offerings, such as sushi and sashimi, bluefin tuna can be quite good.

TUNNY, LITTLE (*EUTHYNNUS ALLETTERATUS*), FALSE ALBACORE*

N.C. Citation Size: Live release of fish thirty-four inches or longer.

Availability: October–November.

Coloration: This fish is bluish-green, with three to five broken, dark wavy lines on its dorsal side. These lines do not extend below its lateral line. The belly is white and is devoid of any markings. There are three to seven dark spots between the pectoral and pelvic fins.

Distinctive Features: The little tunny has a robust, fusiform body that is designed for powerful swimming. For its size, this fish has a relatively large mouth, with the lower jaw slightly protruding past the upper jaw. Scales are lacking on the body except for a patch just behind the head. The caudal fin is markedly lunate, and the caudal peduncle is narrow and has two short keels on either side. The first dorsal fin has high anterior spines that decrease in height toward the small second dorsal fin.

Diet: This species is an opportunistic predator that feeds on fish (herring, sardines, and bay anchovies), squids, and crustaceans.

*At times, the little tunny is mistaken for the Atlantic bonito. While both are members of the tuna family (Scombridae), the bonito differs in appearance from the little tunny in having a low, long dorsal fin; seven to twelve dark, oblique stripes on its back; a slightly compressed body; and very prominent teeth. Bonito can be caught by trolling small spoons, casting lures to breaking fish, and tossing flies. Anglers mainly target them at dawn and dusk when they are particularly active.

Biology & Behavior: This fish is built for speed; estimates have placed its top speed at about forty miles per hour, twice as fast as a bluefish. The combination of a smooth, fusiform body, pectoral fins that fold into small grooves, caudal finlets, and a lunate caudal fin beating at thirty cycles per second makes it ideally suited for eye-blurring bursts of speed. The little tunny is a schooling species, often forming large, elliptical (two miles on the long axis) schools. These schools migrate northward through the coastal waters in the spring and southward in the fall and winter. As befits its name, the little tunny is one of smallest members of the Scombridae family, which includes yellowfin and bluefin tuna, with an average size of up to thirty-two inches and weights up to twenty pounds. Females, upon reaching sexual maturity when between ten and fifteen inches in length, will release almost 2 million eggs in multiple batches into the pelagic waters off the continental shelf. Each egg contains a single droplet of oil that adds to its buoyancy.

Angling Techniques: Many of these fish can be caught quite close to shore because they congregate in areas (points, inlets, rips) where they can trap and ambush prey. As the little tunny slash and shred the bait, the presence of birds, hovering overhead to pick up morsels, will tip you off to the exact location of these fish. Motor up carefully, ease around the school, and cast your offering into the bait. While many anglers use small, shiny lures on medium-sized spinning or bait-casting outfits to catch little tunny, fly-fishing for these marine torpedoes has become quite popular in North Carolina. Your arsenal should include a nine- to ten-weight fly rod, an intermediate sinking line to get down to the action, and a reel that can withstand the initial blistering run of a hooked fish. At the business end of a ten- to fifteen-pound leader, the flies of choice include epoxy minnows (Surf Candy), Clouser minnows, deceivers, and poppers.

Angling Tip: In the heat of the action, anglers often make the mistake of retrieving their lure or fly too rapidly—slow down to have more consistent hookups.

Table Fare: Poor.

WEAKFISH (*CYNOSCION REGALIS*), GRAY TROUT

N.C. Citation Size: Five pounds or live release of fish twenty-four inches or longer.

Availability: April–November.

Coloration: The back is dark olive or blue-green, and the sides are covered in shades of blue, purple, lavender, gold, or copper. Irregular rows of vaguely outlined spots appear above the lateral line but do not extend onto the tail as they do with speckled trout.

Distinctive Features: Two prominent canine teeth protrude from the upper jaw, with narrow bands of teeth on the sides of the upper and lower jaw. Called a weakfish because it has soft mouth tissue that is easily torn by hooks. The body is slim, about four times as long as deep. The first dorsal fin is triangular, originating a little behind the pectoral fin, but the second dorsal fin is much longer than the first, extending almost to the base of the tail.

Diet: Shrimp, large zooplankton, crabs, and small fish (anchovies, herring, and silversides).

Biology & Behavior: Although found from Nova Scotia to Florida, weakfish are most abundant from Long Island to North Carolina. North of Cape Hatteras, the species displays a spring and summer migration northward and inshore, then moves southward and offshore during the fall and winter. They travel in schools that may contain hundreds of fish.

In estuaries, adult trout are usually found in shallow water near the periphery of eelgrass beds and over sandy bottoms. Since they can tolerate salinities as low as ten parts per thousand, weakfish will also reside in salt marsh creeks and river mouths, but they avoid fresh water. In the surf zone, they will congregate in sloughs and channels.

During the breeding season, male weakfish produce a variety of sounds: deep thumps like a drumbeat and bursts of high-pitched croaking. They can also generate loud knocks and crackles when they sense danger or are threatened. The larvae and juveniles grow rapidly, with maturity reached at age one to two. In North Carolina waters, weakfish attain weights between one and four pounds, which are lower than their Chesapeake and Delaware Bay counterparts (four to seven pounds).

Angling Techniques: Many sound anglers use natural baits, like shrimp, fished on a standard two-hook bottom rig to catch weakfish, but leadhead jigs adorned with artificial baits or curly plastic tails catch their fair share of these fish. However, some anglers prefer jigging spoons or lures (two- to three-ounce Stingsilvers are popular). The lure is dropped to the bottom and then reeled up a couple of turns. The angler slowly bounces the lure up and down with a gentle sweep of the rod tip. Many strikes occur as the lure falls, so tension on the line is paramount for success. Because of their varied diet, weakfish forage at various levels, so work the water column or look for concentrations of suspended bait on your depth finder.

Angling Tip: Because of the weakfish's thin mouth membranes, anglers should avoid a hard hook-set; instead they should reel steadily at the first indication of a strike.

Table Fare: Excellent. Fillets are generally broiled or grilled.

Offshore &
Wreck Species

10

AMBERJACK, GREATER (*SERIOLA DUMERILI*)

N.C. Citation Size: Fifty pounds or live release of fish fifty inches or longer.

Availability: June–September.

Coloration: The amberjack has a brownish dorsal side, shading to a silvery white ventral side. There is a dark amber stripe running from the nose to just in front of the dorsal fin. This stripe becomes more defined during periods of feeding or excitement.

Distinctive Features: The amberjack has a slender, elongated body with a short, pointed head and a large mouth. The second dorsal fin is much longer than the first, and the caudal fin is broadly forked.

Diet: Amberjacks are opportunistic predators that feed on benthic and pelagic fish as well as squid and crustaceans.

Biology & Behavior: Found in the offshore region at depths ranging from 60 to 240 feet, the amberjack, like many of the species in this chapter, is quite comfortable with bottom structure—rocky ledges, reefs, and wrecks. Amberjack congregate in schools when they are young, but this schooling tendency decreases as the fish grow older; big amberjacks are primarily solitary. Evidence suggests that migration and spawning are linked, with the latter occurring from March through June over reefs and wrecks.

Research has linked large amberjacks to ciguatera poisoning in humans within certain areas of its range. (I personally do not know of any cases in North Carolina.) Ciguatera poisoning is caused by the accumulation of ciguatoxins, which are produced by marine dinoflagellates (a planktonic organism) in the flesh of subtropical, apex predators.

Angling Techniques: While slow trolling over structure will catch amberjack, the preferred method is dropping live bait—grunt, pinfish, and croaker—to the amberjack's lair. Go with the biggest bait possible to target the largest fish. Hook the bait behind the dorsal fin and under the spine to avoid injuring the fish.

The use of Butterfly jigs, named for their butterflylike fluttering and darting action, has also become popular off Hatteras. These deepwater jigs are unlike traditional jigs that sport a hook at their base; instead they have two stingerlike hooks at the top of the lure. This configuration allows the lure to flutter naturally and has the added advantage of not hanging up when jigging heavy structure. The jigs are fished on spinning outfits loaded with braid.

Since amberjacks are extremely powerful fish, anglers employ heavy tackle to wrestle these brutes up from the depths. A large conventional reel that is spooled with sixty- to eighty-pound line and seated on a six- to seven-foot rod is ideal for providing pressure on the fish and turning the fish's head away from the structure. The terminal tackle should include enough weight to reach the bottom, from eight to sixteen ounces; a 90- to 130-pound leader; and a 9/0 to 11/0 circle or J-hook.

Angling Tip: When attempting to wrench these big fish to the boat may seem the ideal time to employ the traditional "pump-and-reel" technique, aggressive cranking of the reel is the key to success.

Table Fare: Good. The white meat has a firm texture and mild flavor. The amberjack is an extra-lean fish.

BASS, BLACK SEA (*CENTROPRISTIS STRIATA*)

N.C. Citation Size: Four pounds.

Availability: Year-round, but fish move to deeper water during the winter.

Coloration: Like most fish that live on rocky bottoms, black sea bass vary widely in color, ranging from smoky gray to dusky brown to blue-black. The dorsal fin has rows and stripes of white on black.

Distinctive Features: The body is moderately stout, with a relatively high back. Adult fish are nicknamed "humpback" bass because as they grow, they tend to develop fatty bumps just behind their flat-topped head, which has a moderately pointed snout. The spiny and soft ray portions of the dorsal fin are continuous, so that it is only one long fin instead of two short, separate fins. The pectoral fins are quite long, extending almost to the anal fin, and the pelvic fin is also very large.

Diet: Black sea bass feed on a wide variety of fish, mollusks, and worms. Crustaceans, particularly crabs and shrimp, are important food items for all sizes.

Biology & Behavior: Though smaller specimens can be found inshore, the larger fish congregate offshore, where they prefer rocky ledges and a bottom with rubble or some other type of structure. Although they do not travel in schools, sea bass do form large groups around these structures. With their eyes set high on their head, they are excellent visual feeders, seeking out prey during daylight hours. They rely on strong currents and their large mouths to capture their food. Black sea bass can produce a variety of sounds that include weak grunts and thumps; the latter appear to be associated with escape maneuvers and competitive feeding.

Spawning occurs from the middle of May to the end of June. During this period, males develop a hump on their forehead, and the edges of their fins become strikingly cobalt blue. Most offspring are females but change to males when they reach ten inches in length.

Angling Techniques: Target wrecks, reefs, and rough bottom in depths between fifty and eighty feet because these features usually hold the most fish. A two-hook bottom rig with cut bait will suffice in putting fish in the boat. Six to eight ounces of lead is generally sufficient to hold near the structure, and braided line is recommended when fishing deep structure. Since black sea bass are very aggressive feeders, there is no need for assertive hook-setting.

Angling Tip: A single-hook rig with a whole squid for bait is effective in catching bigger fish.

Table Fare: Excellent. The tender and flaky white meat has a subtle flavor that can be enhanced with various spices.

DOLPHIN (*CORYPHAENA HIPPURUS*), MAHI MAHI

N.C. Citation Size: Thirty-five pounds.

Availability: June–September.

Coloration: A dolphin is arguably one of the most colorful pelagic fish, with iridescent greens and blues along its back, changing laterally through a green-gold-yellow spectrum along its flanks, and silvery white or yellow on the belly. However, when the fish is excited, its color may change markedly in an instant, making description difficult. Unfortunately, upon boating, a dolphin fades to a uniform silvery color.

Distinctive Features: Dolphin have an elongated, streamlined body that tapers sharply from head to tail. A single soft dorsal fin extends the length of its body. The caudal fin is strongly forked. Adult males, with their high, blunt forehead, are easily distinguishable from their female counterparts.

Diet: Dolphin are not selective with regard to their prey, although their diet changes as they grow. They feed on small pelagic fish (flying fish, sargassum fish, and triggerfish), juveniles of large pelagic fish (tunas and mackerels), and invertebrates (cephalopods and crabs).

Biology & Behavior: Preferring warm water (78° to 85°F), dolphin are found consistently from the Caribbean basin to the Gulf Stream waters off the North Carolina coast and as far north as New York. Though their migratory habits in the Gulf Stream are a mystery, biologists have hypothesized that they respond to seasonal changes in water temperature in search of more productive food sources. Large males (bulls) and females (cows) generally travel alone or in pairs, and small fish (schoolies or "bailers") travel in schools ranging in size from a few to several dozen. (The number of fish in a school is a function of the size of the individual dolphin.) The largest fish, twenty-five pounds and above, generally show up in the Gulf Stream by early May.

A dolphin is one of fastest-growing game fish in the ocean, increasing in size at a rate of almost one inch in length per week. By the end of its first year, a dolphin may be four feet long and weigh forty pounds. Dolphin spawn in the Gulf Stream from spring through fall, and they may reproduce multiple times per year. Though a dolphin may reach sexual maturity as early as three months, the maximum life span is only four years. Though dolphin generally cruise anywhere within the sunlit layer of the ocean, between the surface and 100 feet deep, anglers often encounter them just below the surface. In fact, they are probably the most surface-oriented of the offshore fish, using their keen sight to forage throughout the day. In the Gulf Stream, dolphin often congregate around floating objects, driftwood, and weed lines, which are thought to offer a variety of advantages: food, protection, and shade.

Angling Techniques: Trolling ballyhoo on twenty- to fifty-pound tackle along sargassum weed lines may elicit a strike from a big dolphin ("gaffer"). When a school of small dolphin is located, generally just below a clump of sargassum or other floating objects (buoys, logs, crates), it's time to switch from trolling to fishing from a drifting boat. The mate chums the water to hold the interest of the school, and anglers use cut bait, such as squid, on light tackle. With reels in free spool, the baited hooks are played out behind the boat. Dolphin are aggressive feeders and will soon engulf the offering, literally hooking themselves. Setting the hook will only have the undesired effect of pulling the bait away from the fish. With the reel now in gear, the angler can control the fish until it can be "bailed" into the fish box by the mate.

Angling Tip: In order to hold the attention of the school, one hooked dolphin should be kept in the water.

Table Fare: Excellent. The firm, white flesh is amenable to a variety of cooking techniques, such as grilling and baking.

HIND, SPECKLED (*EPINEPHELUS DRUMMONDHAYI*), STRAWBERRY GROUPER*

N.C. Citation Size: Twenty pounds.

Availability: May–September.

Coloration: This species is one of the most distinctively colored groupers that can be caught in North Carolina waters. The name stems from the profusion of tiny white spots that cover the deep reddish-brown head, body, and fins.

Distinctive Features: Similar to other groupers, the speckled hind has an oblong-shaped body. The head is long, with a large mouth that has a protruding lower jaw. The caudal fin is broad, with a slightly rounded edge.

Diet: Food items include fish, crabs, shrimps, and mollusks that inhabit hard-bottomed areas.

*Speckled hind is only one species from the family Serranidae that includes other North Carolina groupers: gag, warsaw, black, misty, red, and snowy. Techniques are similar for catching all groupers, but tackle may have to be beefed up for the bigger and deeper grouper, like the warsaw that can reach a weight of more than 300 pounds and resides at depths of 350 to 650 feet.

The speckled hind and warsaw groupers are considered to be at very low population levels and extremely vulnerable to overfishing. In December 2009, the South Atlantic Fishery Management Council passed Snapper Grouper Amendment 17b, which would include a deepwater closure (240 feet seaward) for these species. The amendment is under review by the Department of Commerce.

Biology & Behavior: Speckled hind will hold near low- and high-profile ledges, roughly 150 to 300 feet deep. Off the North Carolina coast, they are found inshore of the deepwater reef fish: tilefish, snowy, warsaw, and yellowedge groupers. The speckled hind is not built for prolonged swimming; it is content to lie in wait and ambush its prey. Upon capturing prey, the species engulfs it whole. The fish opens its mouth and extends its gill covers rapidly to draw in a current of water, literally sucking in the food.

Similar to other groupers, the speckled hind is a protogynous hermaphrodite: it begins life as a female, but after a few years of spawning as a female, it becomes a functional male. The speckled hind is solitary and highly territorial of its piece of the ocean floor.

Angling Techniques: There are two basic approaches to grouper fishing: straight bottom-fishing and free-lining live bait. (See the section on bottom-fishing rigs in Chapter 7.) In either case, since grouper are the quintessential bottom fish, which are quite comfortable holing up in rocks or wrecks, fish your baits, whether live or dead, right on the bottom. If the bait is positioned correctly, the grouper will swim out from its cover, grab the bait, and head back to its lair. How you handle the strike is the critical factor in grouper fishing. Don't abruptly jerk the rod when the fish takes the bait; instead, reel as fast as you can to set the hook. Once the fish is off the bottom, keep applying pressure by steadily reeling. If the fish makes a run in an attempt to get back to the rock, don't abruptly lift the rod, because this action could result in tearing the fish free from the hook. Let the reel's drag and bowed rod fight the grouper. In the unfortunate case that the grouper succeeds in reaching cover, give the fish some slack, rather than breaking the rig off, which at times will deceive the grouper into swimming out from its structure.

Angling Tip: Don't attempt to lift the fish out of the water with the rod. Gaff it in the mouth so you don't lose it after your hard-fought battle.

Table Fare: Good. The flesh of this grouper is snowy white and has a delicate flavor.

MARLIN, BLUE (*MAKAIRA NIGICANS*)

N.C. Citation Size: Live release of fish, regardless of size.

Availability: June–September.

Coloration: The body is dark blue dorsally, shading to a silvery white ventrally. Fifteen light blue or lavender vertical stripes mark the sides.

Distinctive Features: Its large size distinguishes the adult blue marlin from its smaller counterparts, the sailfish and the white marlin. A streamlined and fusiform body, lateral keels on the caudal peduncle, and a lunate caudal fin contribute to making this fish a powerful swimmer of great stamina and speed. A large bill extends from the upper jaw. (The genus name *Makaira* is derived from the Latin *machaera*, which means "sword.") The first dorsal fin is high and pointed in the front but slopes steeply toward the rear. The pectoral fins are pointed, never rigid, and can be folded completely against the sides.

Diet: Without a doubt, blue marlin are the apex predators of the Gulf Stream's food chain. Blue marlin prey upon pelagic fish such as mackerel, dolphin, and tuna with considerable ease. (A forty-pound dolphin is a small meal for a blue marlin.) In fact, they are omnivores, feeding on any organism they can catch.

Biology & Behavior: Blue marlin are solitary hunters that feed primarily, though not exclusively, during the day near the surface. Because they require copious amounts of prey, they are often found near underwater structures—canyons and drop-offs—that attract baitfish. (The Big Rock Blue Marlin Tournament, one of the premier billfish tournaments along the east coast, derives its name from a structure on the edge of North Carolina's continental shelf known as the "Big Rock." The Big Rock—which is not a rock at all—is a compilation of plateaus, peaks, and ledges covering approximately nine square miles of ocean floor. Blue marlin are attracted to this area because it is rich in small fish that seek shelter within this structure.) As consummate predators, marlin rely strongly on visual perception. The optical sensitivity of a marlin's eye is high because it contains many light receptors. These large, optically sensitive eyes may also help the marlin to see while swimming at high speeds in search of prey.

The life history, including oceanic migration and spawning, of the species is poorly understood. But blue marlin are the premier big-game species, noted for their size, strength, long runs, and leaping ability. The largest fish are females, with males rarely topping 300 pounds. The blue marlin off the Outer Banks range from 250 to 400 pounds, but bigger fish, topping 1,000 pounds, inhabit these waters.

Angling Techniques: Troll either large (twelve to sixteen inches), brightly colored lures or bait (large ballyhoo, whole mackerel) on heavy tackle (eighty-pound outfits) over deep water where there is considerable structure or current to attract and hold bait. (Hatteras-based captain Steve Coulter won the 2009 Big Rock Blue Marlin Tournament when a 466-pound blue marlin inhaled his mackerel rig.) Keep in mind that blue marlin are scarce when and where baitfish are limited.

After hookup, a marlin will either jump or sound, often taking several hundred yards of line from the reel. In the latter case, it is imperative to retrieve as much line as possible to minimize the effect of water friction, which could cause line failure. Keep the heat on the fish and do not relax, assuming the fish will soon tire. The time to relax is when the fish is taking line, but the instant the marlin stops peeling line off the reel, switch into a cranking mode. During the fight, a marlin may suddenly swim toward the boat, resulting in significant slack or bow in the line dragging through the water. Be prepared to reel as quickly as possible to tighten the line. Fighting a marlin involves teamwork, from the captain maneuvering the boat to follow the fish to the angler working the fish to the mate controlling a "hot" marlin near the boat. If all of these activities come off smoothly, most marlin, even big ones, can be caught in less than an hour. A marlin does not have to be boated to score a catch; once the mate touches the leader, the fish is officially recognized as caught.

Angling Tip: If light conditions in the water are far from optimal or fish are not picking up visual cues, marlin may rely on mechanical perception—sensitivity to vibrations in the water through use of their lateral line. Try trolling noisy lures that stir up the water surface by creating an enormous bubble trail in their wake.

Table Fare: Not applicable.

MARLIN, WHITE (*KAJIKIA ALBIDUS*)

N.C. Citation Size: Live release of fish, regardless of size.

Availability: June–October.

Coloration: The body is dark blue on top and silvery white on the bottom, with a brown spot on the lower part. In some specimens, many rows of faint white lines run the length of the body. As with many billfish, coloration changes in response to the level of excitement. In particular, its pectoral fins glow an electric blue when the marlin charges into the baits.

Distinctive Features: This billfish has a long, moderately fusiform body. The upper jaw forms a bill that is long and slender compared with other billfish. The first dorsal fin extends almost the entire length of the body and has a large, rounded lobe that tapers abruptly at approximately the twelfth dorsal fin ray. The pelvic and anal fins are also rounded at their tips. The caudal fin has a lunate form.

Diet: Like most billfish, white marlin are apex predators that feed at the top of the food chain. However, lacking teeth, as do all billfish, white marlin are limited to animals that can be swallowed whole. Important prey items are squid and bony fishes, especially flying fish, mackerel, and small dolphin.

Biology & Behavior: White marlin favor deep, blue tropical water and warm temperatures (above 75°F) and generally cruise the waters above the thermocline. As with blue marlin, these fish are sight-oriented feeders, often frequenting regions with distinctive benthic features, such as drop-offs, canyons, and shoals. White marlin do not travel in large schools, preferring to swim alone or in small groups. While swimming or searching for prey, they may exhibit a display known as "tailing"—the dorsal lobe of the caudal fin is visible above the water surface. Females are larger than their male counterparts and live longer, reaching an age of twenty-five to thirty years.

Angling Techniques: Put away the 50- and 80-pound outfits and break out the 20s and 30s. In other words, lighten up. Light tackle will handle white marlin just fine, bring out the best of this acrobatic fish, and are less tiring to handle. Using these light outfits, anglers primarily troll small dead baits. Ballyhoo fished either naked or rigged with a plastic covering is the bait of choice. Many anglers prefer unadorned ballyhoo because it is easier for the marlin to eat; the marlin positions the bait headfirst in its mouth before swallowing it. Since you will be pulling small ballyhoo, also down-size your leaders; lighter leaders (80- to 100-pound test) allow the small baits to swim better. You can rig the ballyhoo to either swim or skip. While both presentations are productive, a skipping ballyhoo may draw a lot more interest and can lead to an exciting encounter with a marlin that is intent on chasing down the bait.

You can't catch marlin until you lure them into your trolling spread. To attract their attention, pull a hookless teaser close to the boat. A proven white marlin teaser is a squid daisy chain. When the fish comes up to the teaser, pull it from the fish and be ready to "feed" the marlin or "drop back." Dropping back is a technique that should be mastered by all serious white marlin anglers and, in principle, involves letting the bait fall back behind the boat as naturally as possible. With the reel out of gear, clicker off, and thumb off the spool, let the line fall from your rod tip into the water for about five seconds. If you perform this movement smoothly, the marlin will eat the bait, with no resistance from you, and you will achieve a solid hookup. If the hookup is missed, and you still have an intact ballyhoo, crank it back up to the surface, hold it there, and be ready to drop back again. A marlin often stuns bait with its bill, so if it strikes with its bill, a dropped-back bait will simulate the stunned prey.

An increasing trend in white marlin (as well as sailfish) fishing is the use of circle hooks. Dropping back J-hooks results in a high proportion of deep hooking, which greatly increases mortality.

Angling Tip: Stay alert. You need to watch the baits and the teaser.
Usually you will spot a white marlin before it reaches the bait,
and you need to have the rod in hand before the fish eats the bait.
Your odds for a hookup increase if you are in control of the situa-
tion. If the fish beats you to the rod, the advantage goes to the fish.
Table Fare: Not applicable.

PORGY, RED (*PAGRUS PAGRUS*), SILVER SNAPPER, PINK SNAPPER

N.C. Citation Size: Four pounds.

Availability: April–July.

Coloration: This fish is named for the rosy tint to its fins and topside. The red porgy has a silvery white underside, and rows of small blue spots dot the body.

Distinctive Features: This fish is medium-sized and has a compact, humpbacked body. It has a large head with a pronounced sloping forehead, large eyes highlighted by blue streaks, and grinding, molarlike teeth. The dorsal fins, which have twelve spines and one soft ray, stretch almost the entire length of the back. Males are significantly larger than females.

Diet: Red porgy are carnivores, feeding on a host of marine animals found on the ocean bottom. Their strong teeth enable them to eat crabs, snails, and sea urchins as well as worms and small fish.

Biology & Behavior: The red porgy belongs to the family Sparidae, or porgies, which also includes pinfish, scup, and sheepshead. This species is a snapperlike fish but has a distinctively different life history. All porgies begin their lives as females, but over time they change sex to become functional males. Researchers have documented that the sex change commonly occurs at age three, with some individuals not changing until almost seven years old. The exact factors, whether environmental and/or social, that trigger this sexual transformation have remained elusive to fishery biologists. Red porgy spawn during winter and early spring, with females producing eggs throughout the spawning season. These eggs are released into the surrounding waters, where they are externally fertilized. The hatched larvae and juveniles grow fast during their first four years, and adults are known to live up to eighteen years.

Angling Techniques: Though red porgy can be found in both shallow and deep waters, the larger fish are generally caught from a boat over hard-bottomed areas in deeper water. As with groupers and snappers, finding underwater structures, such as offshore reefs, is the key to successful porgy fishing. In fact, red porgy are often caught by anglers targeting grouper. Although porgy are hard-fighting, tenacious adversaries, light tackle will tame these bantamweights. Medium-action spinning rods and reels with eight- to ten-pound line are an ideal setup if the water is not too deep. Most porgy fishermen employ a multiple-hook rig to increase the chance of a hookup, since porgy become more aggressive when competing for food. Use a bank sinker heavy enough to hold bottom. Popular baits include worms, squid, and clams. Worms should be threaded on the hook, with a small piece extending beyond the hook. Clams, having very soft flesh, should be threaded onto the hook multiple times, but tough squid strips only need to be threaded once or twice.

Porgies respond quickly to the scent of chum in the water. A chum pot, holding a slab of frozen chum, is lowered to the bottom, generally up-current of where the baits are presented.

Angling Tip: When the bite is on, use the smallest pieces of bait to take advantage of the increased aggressiveness of these feisty fish.

Table Fare: Good. White fillets have a light to mild flavor.

SAILFISH (*ISTIOPHORUS PLATYPTERUS*)

N.C. Citation Size: Live release of fish, regardless of size.

Availability: June–October.

Coloration: The body is dark blue dorsally and silvery white ventrally. The coloration of a sailfish may change markedly, depending on its level of excitement. The blue bars, about twenty of them, on each side can dramatically increase in intensity and contrast.

Distinctive Features: The sailfish get its name from its characteristic sail-like dorsal fin that runs most of the length of the body and is much taller than the width of the body. The upper jaw, which is approximately twice the length of the lower jaw, becomes a long bill. The pelvic fins are also long, extending almost back to the first anal fin. Like all billfish, sailfish have a lunate caudal fin.

Diet: Flying fish and squid, preferring mackerels, jacks, and other fish that swim near the surface.

Biology & Behavior: Sailfish grow fast, attaining lengths of four to five feet in a single year, but they are still one of the smallest members of the billfish family. A typical Outer Banks sail will weigh between thirty and fifty pounds, with the biggest weighing in at about eighty pounds.

They usually travel alone or in small groups, but the individuals may work together to concentrate prey into small circles. One of the fastest fish in the ocean, sailfish can reach speeds approaching seventy miles per hour because of their streamlined body, water-slicing bill, and a pair of grooves along the ventral side of the body, into which the pelvic fins can be recessed. Sailfish usually keep their sails folded down when swimming, but if threatened or when herding prey, they may raise their sail, giving the impression of a much larger organism.

Like many other pelagic species, sailfish are prodigious spawners, releasing 4 to 5 million eggs per spawning event. Fertilization occurs in the open water, where the eggs drift with the currents until hatching. The larvae grow rapidly, reportedly reaching four to five feet in length in one year.

Angling Techniques: First of all, anglers should realize that North Carolina's offshore waters are not a hotbed for sailfish angling. That said, slow trolling (five to six knots) dead baits on light tackle (twenty-pound) is the norm for most charter boat captains. However, the technique requires some finesse. Sailfish are very sensitive when they pick up the bait; any resistance, such as too tight a drag on the reel, may pull the bait from the fish's mouth. Generally, the reel is put in free spool to allow the fish to turn and run with the bait. The angler allows the fish to eat for at least five to ten seconds before engaging the reel.

Angling Tip: If the sailfish misses the bait or is skittish, anglers may often feed the fish—dropping the bait back to the fish.

Table Fare: Not applicable.

SHARK, SHORTFIN MAKO (*ISURUS OXYRINCHUS*)

N.C. Citation Size: 150 pounds or live release of shark eighty inches or longer.

Availability: Year-round in the Gulf Stream.

Coloration: One of the most colorful of all shark species, the mako is brilliant metallic blue dorsally and white ventrally. A distinct line of demarcation separates the blue and white on the body. The underside of the snout is also white. Color changes with size; larger specimens possess darker color that extends onto parts of the body that are white in smaller individuals.

Distinctive Features: The shortfin mako has a streamlined body with a sharply pointed snout, well-developed black eyes, five large gill slits, a large keel on the caudal peduncle, and a pronounced lunate tail. Its large dorsal fin begins just behind its winglike pectoral fins. The mako's bowl-shaped mouth is lined, both top and bottom, with triangle-shaped, razor-sharp teeth that lack cusps or serrations found in other shark species.

Diet: As a top-level predator, a mako feeds on other fast-moving pelagic fish, including tuna, swordfish, and other sharks.

Biology and Behavior: Though these sharks prefer warm water (63°–68°F), they are able to adapt to temperatures in the lower fifties because of their ability to elevate their body temperature. In this sense, makos are warm-blooded, meaning that heat in their blood is not lost to the environment through the gills. Tag-and-release studies have shown that this species migrates seasonally to warmer waters.

As in other shark species, females are larger than males, reaching weights of more than 700 pounds. These sharks have a rapid growth rate—twice as fast as some other shark species. Males mature when they reach a length approaching seven feet; females, about nine feet. Both sexes reach sexual maturity before age six. Shortfin mako development is ovoviviparous: the fertilized eggs hatch inside the female after a fifteen- to eighteen-month gestation period. Once the young are hatched into the uterus, cannibalism—known as oophagy—may occur that results in the ingestion of less-developed eggs by a fetus that is more developed. The resulting litters range from eight to ten pups.

The shortfin mako is the fastest shark, reaching sustained speeds of over twenty-five miles per hour, and has been known to travel more than 1,300 miles in little over a month.

Angling Techniques: While the mako doesn't get the attention that other offshore species, like marlin and tuna, garner from anglers, it is one species that you should consider adding to your fishing portfolio. The mako is known for its power, bursts of speed, and acrobatic jumps, and the author Zane Grey propelled it into big-game fishing status.

Probably the most important aspect of mako fishing is to set up a good chum slick on a thermal break. Large quantities of ground-up menhaden produce a very oily slick that floats on the surface. The tackle is fairly basic: a fifty-pound outfit that is spooled with about 500 yards of monofilament, a ten- to fifteen-foot wire leader, and hook sizes in the 6/0 to 10/0 range. Depending on what is available, whole baits, preferably live, or long strip baits (twelve to fifteen inches) of bluefish, mackerel, or tuna can be used. The baits are often floated behind the stern at different depths, using balloons, or if a mako appears in the slick, the baits can simply be deployed overboard. With the chum slick streaming out behind the boat, all that remains is to start drifting.

Upon hookup, a mako should be fought differently from other fish. The goal is not to bring the fish as quickly back to the boat as possible, as you would with a tuna. Two key points should be considered in fighting the mako: initially keep the mako away from the boat, since you don't want a spirited mako jumping into the boat, and keep pressure on it until it tires, before bringing it alongside the boat.

Angling Tip: Shark fishing is inherently dangerous because of the nature of the beast. But a mako's size, aggressiveness, and unpredictably make it particularly worthy of your respect. There are plenty of mako disaster stories of wrecked cockpits and injuries. If you are a novice to this type of fishing, go with an experienced captain.

Table Fare: Excellent. Mako steaks work great on the grill.

SNAPPER, RED (*LUTJANUS CAMPECHANUS*)*

N.C. Citation Size: Ten pounds.
Availability: April–November.
Coloration: The body and fins are pinkish red, with the most vivid coloration on the back and lightening to a white underside. Juvenile snapper, less than fourteen inches, have a dark spot on their sides below the dorsal fins.
Distinctive Features: Like other snappers, the red snapper features a laterally compressed body that tapers from a large head to a truncated caudal fin. The first and second dorsal fins are continuous but marked with a slight notch between them. The species has long pectoral fins. The teeth of the red snapper are short, sharp, and densely packed. There are some upper canine teeth, but they lack the size found in mutton and mangrove snappers.

*Other North Carolina snappers include vermillion and the occasional cubera and mutton snapper, which are found primarily in a tropical habitat. While cubera and mutton snappers have habits and characteristics similar to those of a red snapper, the vermillion snapper aren't big, so lighter tackle, even spinning reels, can be used quite effectively. Standard two-hook bottom rigs are fine, but since vermillion snappers have small mouths, you need to use small hooks. Since these fish are not true bottom-feeders, dropping your rig all the way to the bottom will only occasionally entice a snapper to your bait. Instead, stop your rig about halfway to the bottom or reel it back to midwater depths after it has reached the bottom.

169

Diet: Throughout its life span, the red snapper is a carnivore. Juveniles feed on tiny zooplankton, such as copepods, but as they mature, they switch to larger prey, including shrimp, squid, octopus, and fish.

Biology & Behavior: Adult red snappers reside at depths from 30 to 500 feet, with the older, larger fish seeking out the cooler, deeper sites. The preferred habitats for the adults are ledges, rocky bottoms, artificial reefs, and wrecks. Around these structures, red snappers, like most other snappers, will form large schools of similar-sized fish. Being structure oriented, red snapper forage on or near the bottom. With regard to reproduction, all snappers are oviparous, producing eggs that hatch outside the female's body; they are also prolific spawners, producing hundreds of thousands of eggs during a single spawning event. Estimated maximum age is forty to fifty years.

Angling Techniques: As in many angling scenarios, live bait, such as menhaden and pinfish, is best for red snapper fishing. In lieu of the above offering, the choice of many anglers is cut bait, such as squid, sliced into strips or presented whole. The tackle used depends on the location and depth that you have selected to bottom-fish. In deeper waters, especially with strong current flow, heavy tackle is needed to pull these tenacious fish from their lairs: stout boat rods (with a sensitive tip to detect the bite), conventional reels (preferably two-speed), and fifty-pound braided line. The terminal tackle is relatively simple: a long leader to avoid spooking the fish, a large circle hook (5/0 to 6/0), and a heavy sinker (one to two pounds) attached to the bottom of the leader.

Another method that has grown in popularity is jigging. Since red snapper are bottom-dwellers, they can be enticed to strike the jigs that are tipped with a squid or fish strip. A long fluorocarbon leader forms the "invisible" link between the braided line and the jig. Depending on water clarity, vary the pace of the jigging; in clear water, move the jig faster, but slow down your strokes when visibility is decreased.

Red snapper fishing is generally most productive from an anchored boat, which allows the onboard anglers to drop their offering consistently over the target zone. With any big red snapper, the first few moments after the hookup can mean the difference between success and failure. After the strike, you need to retrieve as much line as possible before the feisty and determined snapper surges back to its structure.

Angling Tip: Have on hand a good supply of quick-change leaders that can replace the terminal tackle that you will invariably lose on the rocky bottom.

Table Fare: Excellent. Fillets are flaky when cooked, with a very mild and sweet flavor.

TILEFISH, BLUELINE (*CAULOLATILUS MICROPS*), GRAY TILEFISH*

N.C. Citation Size: Eight pounds.

Availability: Year-round, weather permitting.

Coloration: The species is a dull olive-gray overall, with a white belly.

Distinctive Features: The body is stout and elongated and tapers to a broad caudal fin. A tilefish's head is large and has a pointed snout. The lack of a fleshy protuberance just behind its head distinguishes it from the commercially important golden tilefish. The large, toothy mouth has serious crushing power, a necessity for subduing the hard bottom creatures these fish dine on. Both the dorsal and anal fins are long and continuous, extending to the base of the caudal fin.

Diet: Tilefish are bottom-browsers that often feed on crabs, brittle stars, snails, worms, and sea urchins.

Biology & Behavior: The species is a deep-bottom dweller found in depths from 250 to 800 feet, from Virginia to the Campeche Banks of Mexico. It frequents irregular bottoms comprised of troughs and terraces intermingled with sand, mud, or shell substrates. Tilefish commonly share the same habitat with other deepwater species, such as snowy groupers. It has been observed hovering

*Golden tilefish also inhabit the deep ledges and crevices of North Carolina's offshore waters but are considerably larger than the blueline tilefish—reaching weights up to seventy pounds.

near or entering holes under rocks. The species is also known to root headfirst into cone-shaped sand piles, in an attempt to root out its prey. Long lived and slow growing, blueline tilefish may attain lengths of almost three feet in fifteen years.

Angling Technique: Fishing for tilefish is classic deep-drop angling, requiring either large conventional reels or electric reels. The line of choice is braid (50- to 100-pound test) because of its minimum stretch. (The stretch in monofilament negates the angler's ability to detect the bite at great depths.) Sinkers can be as heavy as a few pounds, depending on depth and current. Multiple, large circle hooks are baited with an assortment of baits to accommodate the varied diet of tilefish. Locating the fish is not always easy because of their habit of forming burrows. Odds are that the largest fish caught will be the first, because of their aggressive nature.

Angling Tip: While cut bait is the preference, it needs to be fresh enough to stay on the hook. Frozen bait will be very soft when it thaws, and during the long drop, soft bait will tear from the hook before it reaches the bottom.

Table Fare: Good. The fish has a mild flavor that some have compared to lobster or scallops. Because of the relatively high levels of methyl mercury found in these fish, the Food and Drug Administration has advised pregnant and nursing women to limit their intake of this species. (Methyl mercury seems to be a problem with many long-lived pelagic fish, including king mackerel and tuna.)

TRIGGERFISH, GRAY (*BALISTES CAPRISCUS*)

N.C. Citation Size: Five pounds.

Availability: June–September.

Coloration: Adult fish exhibit a light gray to olive-gray body color, with three faint, broad, and dark blotches on the upper body and white dots on the lower body and fins. Triggerfish often change their coloration to match their surroundings.

Distinctive Features: The body of the gray triggerfish is laterally compressed, with sandpaperlike skin. The species has two dorsal fins, but the first has three sharp spines that can be locked in an erect position. The triggerfish derives its name from the action of the spines. When the second spine is depressed, it acts as a trigger that unlocks the first spine. By erecting and locking its spiny dorsal fin, the triggerfish can position itself immovably in bottom crevices.

Diet: The species, relying on its powerful teeth, dislodges and crushes small mussels, sea urchins, and barnacles.

Biology & Behavior: Triggerfish rely on their dorsal fins for locomotion, flapping the fins back and forth to propel themselves forward. In addition, triggerfish use undulating motions of their dorsal and anal fins to ascend and descend, as well as hover over the bottom searching for food. During the summer months, triggerfish build their nests on the bottom, where they deposit between 50,000 and 100,000 eggs, depending on the size of the female. The adult fish vigorously guard the nests from all intruders, including divers. After hatching, the larvae leave the nests and ascend to the surface, where they commonly associate with sargassum communities. As autumn approaches, the juvenile triggerfish leave the sargassum habitat for a bottom dwelling, such as a reef or rock outcrop.

Angling Techniques: Large triggerfish can be caught over relatively shallow (forty feet) wrecks and reefs. Heavy tackle is not needed; fifteen- to twenty-pound tackle, with enough weight to hold the bottom, will suffice. Triggerfish are notorious nibblers, so you must vigilantly watch your lines to detect a bite. If the line twitches, avoid jerking hard to set the hook because a triggerfish's mouth is bony, and a quick jerk results in the hook sliding across the mouth instead of attaining purchase. The technique is to lift up the rod and apply steady pressure by reeling as quickly as possible. If the fish has the hook in its mouth, and as long as you maintain a tight line, odds are in your favor of landing the fish.

Triggerfish have extremely small mouths, so downsize your hooks (1/0 to 3/0) to accommodate them. Use sharp hooks and keep them sharp. A fish-finder rig baited with almost any kind of bite-sized, cut bait will do the trick. But the tougher the bait the better—make it difficult for the trigger to pilfer the bait. Once you're rigged up, drop your offering to the bottom and quickly take up any slack, so you can detect the strike.

Angling Tip: As opposed to some reef fish, triggerfish often swim up with the bait. All the angler may notice is a lack of weight on the end of the line. If the line goes slack, keep reeling—there may well be a fish on.

Table Fare: Good. The flesh has a very delicate flavor, but the fish's tough skin makes cleaning tricky.

TUNA, YELLOWFIN (*THUNNUS ALBACARES*)*

N.C. Citation Size: Seventy pounds.

Availability: April–May and October–December.

Coloration: The tuna's body is metallic blue on top and shades to a silvery white on the bottom. Both the dorsal and ventral portions of the fish are crossed by parallel vertical lines. A golden or iridescent blue lateral line runs along the entire length of the body, although it is not always prominent. The second dorsal fin and the anal fin are bright yellow.

Distinctive Features: The body is strongly fusiform, deepest under its first dorsal fin and tapering markedly toward the caudal peduncle. Two dorsal fins are present, with the second fin being very long in adults, and these fins become relatively longer in larger individuals. The pectoral fins are also long, reaching back to the second dorsal fin. The caudal peduncle includes seven to ten dorsal and ventral finlets. The caudal fin is lunate.

Diet: Tuna primarily prey on fish, cephalopods, and crustaceans, which they forage indiscriminately.

*Though not targeted as frequently as their yellowfin cousin, blackfin and bigeye tuna are sometimes caught by the offshore fleet. Blackfin tuna is the smallest tuna species, growing to a maximum of three feet in length and weighing forty pounds. In contrast, a bigeye tuna weighs over 100 pounds. Both are caught by trolling, but bigeye tuna often inhabit depths of more than 500 feet, searching for deepwater prey.

Biology & Behavior: The yellowfin tuna is an epipelagic fish, living
in the upper 300 feet of the water column and preferring warm
water. However, since yellowfin are primarily sight-oriented
predators, their feeding tends to occur in the surface layers during
daylight hours. But since fish leave a scent in the water comprised
of the oils, proteins, and amino acids from the slime layer on their
bodies, yellowfin tuna can readily pick up the scent trail and actually track down their prey.

Yellowfin tuna exhibit strong schooling tendencies and swim
together in large schools of same-sized fish. All yellowfin, like
most tuna, are serial spawners, meaning they may release batches
of eggs year-round. Yellowfin are reproductively mature by age
two to three years and may live up to seven years. But by age four,
they may reach weights of over 125 pounds, though their maximum size can be more than 300 pounds.

Angling Techniques: Because of the yellowfin's eclectic diet, anglers
may troll a variety of offerings: naked and skirted (feathered dressing) ballyhoo, artificial squid that are rigged in line (daisy chain)
to act as a teaser, and bird rigs that have a flying fish profile. Since
flying fish—a favorite prey of yellowfin tuna—often propel themselves into the air to avoid predation, some anglers attach the bird
rig to a kite and then manipulate the kite to simulate the flight of
a flying fish.

Due to their schooling nature, yellowfin tuna can often be
spotted feeding on the surface, driving pods of bait upward from
the depths. Look for birds that are wheeling overhead, patiently
waiting to pick up the morsels. Trolling through this melee almost
guarantees multiple hookups.

Upon hookup, yellowfin tuna dive deep and slowly swim in circles beneath the boat. The trick is to retrieve as much line as possible when the fish is on the inward leg of the circle and minimize
yielding line to the fish on the outward leg. Rhythmically pumping the rod is the routine to subdue this pelagic bulldog.

Angling Tip: In the spring (a peak season), yellowfin tuna are found
as close as twenty miles from the coast, so anglers can enjoy six to
seven hours of fishing on an all-day charter.

Table Fare: Good. Tuna steaks are best grilled, but be alert not to
overcook them.

WAHOO (*ACANTHOCYBIUM SOLANDRI*)

N.C. Citation Size: Forty pounds.

Availability: June–September.

Coloration: The body is dark blue along its back, with two dozen wavy, cobalt blue bars running vertically along its sides. The belly and lower body are silvery.

Distinctive Features: The jaws of a wahoo are elongated to form an almost beaklike snout and contain dozens of large, triangular, and finely serrated teeth. The wahoo has a slender, fusiform body. The caudal peduncle is narrow, contains three sets of keels, and tapers to an almost lunate caudal fin.

Diet: Wahoo are almost totally piscivorous, feeding on herrings, jacks, mackerels, flying fish, and many other pelagic species.

Biology & Behavior: Wahoo are generally solitary hunters or found in very small, loosely associated groups and never form large schools like dolphin. Wahoo are capable of attaining swimming speeds of up to fifty miles per hour in short bursts that permit them to pursue, overcome, and capture prey in the blink of any eye. Wahoo grow rapidly and reach lengths of four to five feet, though some have topped out at over eight feet. Both sexes reach maturity between one and two years of age and may live more than five years.

Angling Techniques: As with many other Gulf Stream species, trolling is the main method for targeting wahoo. Rigged natural baits, such as ballyhoo, are fished on wire line outfits at depth, generally using downriggers or planers to position the bait deeper in the water column. The heavy wire line has the advantage of tracking straight behind the boat. But anglers who are knowledgeable about the predatory habits and food preferences of wahoo troll fishlike lures in straight lines at relatively high speeds of over twelve knots. Though the sleek, tapered lures are efficient tools, displacing far less water than flat-head chuggers, the only way to keep them from skipping out of the wake is to employ an in-line sinker (twenty-four to thirty-six ounces) several yards in front of the leader.

Though pelagic in nature, wahoo often associate with structure, such as pinnacles, wrecks, and deep rock piles, where prey are abundant. Troll around these features but be aware that there are times when wahoo will gather around flotsam, much like dolphin.

Angling Tip: Regardless of the method employed to catch wahoo, a wire leader, inserted between the hook and the main line, is good insurance against the impressive dental work of the wahoo.

Table Fare: Excellent. Wahoo steaks are best grilled or pan seared. The Hawaiian word for this fish is *ono*—literally, "good to eat."

After the Catch *Caring for & Cooking Fish*

Keeping a few fish for dinner is, to me and many others, one of the more satisfying aspects of angling along the Outer Banks. In addition to having nutritional value (high in protein, low in saturated fats, and a good source of omega-3), most fish are just tasty, offering a variety of flavors and textures that appeal to even the most discriminating palates. John Hersey, in his book *Blues*, probably summed it up best when he commented about eating a bluefish that had been caught by his fishing companion: "And all the more sweeter because you caught it." If "location, location, location" is the maxim of the real estate establishment, then "freshness, freshness, freshness" should be the mantra of the angling community. A fresh fish shouldn't smell "fishy." If it does, the fish's quality has deteriorated—and that smell will only intensify when you cook the fish.

Upon landing the fish, try to avoid bruising it against hard surfaces, such as boat decks or gunwales, and minimize its exposure to the sun. High summertime temperatures can cause spoilage in less than an hour due to bacterial growth. However, simply throwing your catch into the fish box is also a recipe for disaster, since the high body temperature of the fish will "cook" it internally. In order to maintain optimum quality, fish should be chilled, preferably in ice, immediately after being caught. When ice is in direct contact with fish, it chills the fish by surface heat transfer. Don't scrimp on the ice; the more ice in contact with the fish, the quicker the cooling rate. When possible, use crushed ice, as opposed to block ice, since the greater surface area of crushed ice will mean more of it is in direct contact with the fish.

A brine slush-ice mixture, which is made by adding equal amounts of seawater and ice to the fish box or cooler, is most effective for chilling fish. Though some melting of the ice occurs, the overall temperature of the mixture does not change markedly because of the salt in the seawater. The result is a chilly slush with a temperature of approximately 32°F, but with more surface area in contact with the fish. Care should be taken to ensure that the seawater is clean and not contaminated with dirt, fuel, or oil.

After fishing, clean your catch as soon as possible. While fish tissue is almost sterile, the skin surface and viscera may contain different types of bacteria. In principle, cleaning your catch seems quite simple—either gut the fish or cut fillets from the carcass. Improper technique or rough treatment of the fish can lead to gouges or wounds in the flesh that are pathways for the spread of bacteria, not to mention an unsightly fish that is destined for the dinner table.

If the fish has been chilled properly, cleaning it can be a breeze if you have the right tools. But first you need to answer a few questions: How big is the fish? Are you going to fillet or steak it? Will the skin remain on one side of the fillets? While you may be able to get by with one knife, different cleaning tasks are made easier if you have more than one knife.

Knives come in a variety of blade lengths (six to fourteen inches) and styles (thin, wide, straight, curved, smooth, or serrated). While a serrated knife can be used for carving steaks from a fish, it is mainly for cutting through ribs and bone. For small fish, such as trout or Spanish mackerel, a six- to eight-inch, straight blade is ideal for removing fillets. If the blade is narrow, detail work, like removing the skin from the fillets, is also possible. For larger fish, such as cobia, grouper, or snapper, a long (fourteen-inch) curved blade makes carving easier. In all cases, keep your knives sharp; a dull edge will only make your task harder.

Proper filleting of a fish takes practice. If you want to see filleting taken to its highest level, stop by one of the fish-cleaning stations at the local marinas. Almost on a daily basis, a small crew deftly slices through hundreds of pounds of fish in short order.

Cleaned fish should be washed under cold, running water and patted dry with absorbent paper towels. Avoid immersing fillets in a prolonged freshwater soak that could dilute and reduce flavor and texture. If the fish is not to be consumed within the next few hours, it can be refrigerated in airtight plastic bags. Though the shelf life depends on the species and how well it is cared for, most fish suffer a loss in flavor and texture after two days. Your "fresh" fish is no longer fresh.

If the fish cannot be consumed within a few days of being caught, freezing is a viable option because at subfreezing temperatures illness-causing bacteria cannot grow. The key to maintaining some degree of quality in the fish is in its storage. Fish needs to be wrapped appropriately to minimize exposure to air and to prevent freezer burn. Not all fish equally tolerate long periods of being frozen. Firm, white flesh does better in the freezer than oily fish. The flesh of large bluefish, no matter how carefully prepared before freezing, becomes mushy over time.

The recipe you choose for cooking your fish depends on the time and effort you want to expend. While you might want to impress your guests by cooking blackened redfish—a recipe made famous by Louisiana chef Paul Prudhomme—be aware that if you cook indoors, the smoke from the cast-iron skillet will set off every fire alarm in the house. For cooks even more adventurous, consider the recipes that Mark Kurlansky has recorded in his book *Cod: A Biography of the Fish That Changed the World*. How about fried cod (substitute your favorite Outer Banks fish) head, with lips removed. Stewed codfish tongues, anyone?

If you are on vacation, you probably want to keep the ingredients, preparation, and cooking to a minimum. But you don't have to sacrifice taste when you keep it simple. Fish can be cooked in a variety of ways: poaching, steaming, broiling, baking, grilling, frying, and sautéing.

Poaching is cooking the fish in a flavorful bath, such as wine or fish stock, which adds moisture to the fish and protects its delicate flesh. Poaching works well with firm, white fish. In addition to being relatively quick and easy, steaming rarely results in a dried-out fish because the heat is even and enveloping, allowing sufficient time on your part for any adjustments. Don't be afraid to use the oven even in the summertime. Whether broiled or baked, fish will cook quicker than denser meats or roasts. When broiling, place the fish close to the top heat source and use a greased, heavy-duty cooking sheet for ease in turning and plating the fish. Since baking employs dry heat, thick fillets or whole fish won't dry out quickly. Who doesn't like grilling, whether on a gas grill or using charcoal? There is something primal about cooking over an open flame. The keys to success in both cases are to carefully monitor and control the heat to prevent the fish from burning and drying out. While some chefs make a distinction between sautéing and pan-frying, others use the terms interchangeably. Both methods require the use of a fat, such as oil or butter, but sautéing requires less fat than frying—only enough fat to lightly coat the bottom of the pan. When sautéing, ingredients are cut into pieces or thinly sliced to facilitate fast cooking, but larger pieces are employed in frying and are flipped onto both sides.

Whichever method of cooking you choose, a question that is always out there is How do I tell if my fish is cooked? While experience is a definite plus, and practice can lead to experience, below are some guidelines:

- A general rule of cooking fish is ten minutes per inch.
- A fish changes color as it cooks. Because a fish is high in protein, when the protein is exposed to high heat, the fish tissue will change from translucent to opaque. Fried food will change to a golden brown. Watch how color changes with a heat source from below, as in frying, as opposed to broiling, with the heat source from above.
- A fish's texture changes as it cooks. During cooking, the fish's flesh will become firmer than when it was raw. As the fish cooks, use the flat part of your index finger to test the resiliency of the flesh. Over time, you will be able to determine the right firmness that indicates a cooked fish.
- Remove the fish from the heat source before it is completely cooked. As you plate the fish, the internal, residual heat will continue to cook the fish.

The recipes that I present below, and have tried, are included in this book to give you some idea of how you can cook your Outer Banks catch.

Marinated Grilled Yellowfin Tuna

| ½ cup soy sauce | 2 teaspoons wasabi paste |
| ¼ cup honey | 2–4 tuna steaks |

Combine the soy sauce, honey, and wasabi paste in a bowl and whisk thoroughly together. Add the tuna steaks, assuring that they are covered by the marinade. Cover the bowl and chill in the refrigerator for one to two hours.

I prefer grilling the tuna on a hot, charcoal fire—the hotter the better. Once the coals are glowing, place the steaks on a lightly greased grill. Cook for approximately one to two minutes on all sides. The tuna should be seared on the outside but pink on the inside.

Panko Fried Flounder *Substitutes: croaker, trout, kingfish*

⅛ cup Old Bay seasoning

1 cup flour

4 skinless flounder fillets

2–3 eggs, beaten for an egg wash

1 cup Japanese panko breadcumbs

¼ cup vegetable oil

salt to taste

Mix the Old Bay seasoning into the flour to give the fish a little more "kick" in flavor. Lightly coat the fillets in the flour mixture and dip them in the egg wash. Once the fillets are thoroughly coated, place them in the panko breadcrumbs, making sure there is good adhesion. (You can use plain breadcrumbs, but the panko breadcrumbs give the fish an almost potato chip crispness when fried.)

In a nonstick skillet, heat the oil until it starts to shimmer. Place the fillets in the skillet and cook until both sides are golden brown. Drain on absorbent paper towels and season with salt while fish is still warm.

Broiled Bluefish *Substitute: Spanish mackerel*

4 medium-sized, skinless bluefish fillets

sea salt and ground pepper

¼ cup mayonnaise

3 tablespoons coarse mustard

1 tablespoon Worcestershire sauce

2 tablespoons chili sauce

From the fillets remove the center bloodline (dark band) and season with salt and pepper. In a small bowl, combine the mayonnaise, mustard, Worcestershire, and chili sauces. Paint both sides of the fillets with the mayonnaise coating, which will not only add flavor but hold the fish's moisture in under the high broiling temperatures.

Place the fillets a few inches from the top broiler and broil both sides for about three to four minutes, until the coating is brown and bubbling.

Baked Tomato-Chili Dolphin *Substitutes: grouper, snapper*

4 dolphin fillets	¼ cup green olives
2 limes, juiced	3–6 chilies
4 cloves garlic, minced	2 tablespoons capers
2 tablespoons olive oil	1 teaspoon oregano
3 green onions, chopped	3 tablespoons chopped parsley
6 plum tomatoes, diced	salt and pepper to taste

Marinate the fillets in the lime juice and two garlic cloves for thirty minutes. Add the olive oil to a hot pan and then add the onions and remaining garlic, stirring for about a minute. Add the tomatoes and cook for another minute. To this mixture add the olives, chilies, capers, oregano, and parsley. Let it simmer for a few minutes to blend the flavors. In a baking dish, arrange the fillets and pour the tomato mixture evenly over them. Bake in a 350° oven for twenty minutes. Season to taste with salt and pepper.

Steamed Black Sea Bass *Substitute: Porgy*

1 whole black sea bass (2–3 pounds), scaled, gutted, and fins removed
¼ cup green onions, chopped
¼ cup shallots, chopped
¼ cup fresh ginger, chopped
¼ cup soy sauce
3–4 Chinese cabbage leaves

Make three deep diagonal cuts into both sides of the fish, but not down to the bone. In the slits and the cavity, evenly distribute onions, shallots, and ginger. Place fish in baking dish, pour the soy sauce over it, and marinate for thirty minutes.

Line the steamer with the cabbage leaves and then put fish over leaves. Pour about two cups of water into pan or pot that holds the steamer and bring to a boil. Set the steamer basket over the boiling water, cover, and steam for about ten minutes, or until the fish is fully cooked.

Poached Pompano with Creamy Clam Sauce

Substitute: cobia

4 pompano fillets	¼ stick (2 tablespoons) butter
salt and pepper to taste	1 tablespoon flour
½ cup white wine	½ pint cream
½ cup clam juice	

Salt and pepper the fillets. Place them in a pan and cover with white wine and clam juice. Over medium heat, cook the fish for eight to ten minutes, until fork-tender. Remove the fish and keep it warm. Reduce the poaching liquid under high heat, strain it, and set aside.

In a small pot, melt the butter and slowly add the flour to form a roux. When the roux is at the desired consistency, add the liquid and heat again. Stir the cream into the heated mixture. Pour this sauce over the fillets.

Epilogue

If you fish the waters of the Outer Banks long and hard enough, you're bound to encounter at least one of the following critters. They can at times be quite abundant and a major nuisance—"stealing" bait, breaking off rigs, and inflicting pain to the careless angler. These species are lumped together under the unflattering heading of "trash" fish and are generally perceived to have little or no value compared with the more desired game fish. I have included only five (in no particular ranking) species, because from my experience they have caused the most angst among anglers. But you may already have your personal favorite, such as the oyster toadfish or lizardfish.

Smooth dogfish (*Mustelus canis*): Any little gray or brownish shark that is distinguished from other sharks by its slender body and flat, blunt teeth. Smooth dogfish swim in packs or schools and give birth (pupping) to live young, possibly the basis for the word "dogfish." A scavenger and opportunistic predator, the smooth dogfish feeds on invertebrates, crustaceans, and small fish. These fish can be so thick in the surf that every cast results in a hookup.

Skate, family Rajidae: Skates are cartilaginous fish that have a flattened and broadly rounded body, winglike pectoral fins, and a short, slender, and spineless tail. Skates are bottom-dwellers, partly burying themselves in the nearshore sediments. Because their mouth is located on the underside, they pounce on their prey before consuming it. This feeding style makes them very adept at consuming the bait from an angler's bottom rig. Their developing eggs are encapsulated in amber or black leathery cases (mermaid's purses) that frequently wash up on the beach.

Stingray, family Dasyatidae: Stingrays are cartilaginous marine fish and are related to skates and sharks. Their tail, which is slender and whiplike, possesses one or more venomous spines or barbs that can cause painful and potentially serious wounds to a careless angler. Stingrays, like skates, are bottom-feeders, feasting on worms, crustaceans, and mollusks. As opposed to skates, stingrays give birth to live young. Rays are also considerably larger than skates. The common southern stingray measures more than six feet from wing tip to wing tip and easily tops out at over 100 pounds.

Barracuda (*Sphyraena barracuda*): Though often found on shallow coral reefs, barracuda regularly swim in the offshore waters of North Carolina and, in particular, ride the Gulf Stream northward during the summer. The distinguishing feature of these long, slender fish is their large mouth that has two sets of razor-sharp teeth that make them feared predators of a wide variety of fish. Barracuda locate their prey primarily by sight and use their speed (thirty-six miles per hour) to chase down their food. In search of an easy meal, a barracuda (they generally tend to be solitary) will attack the trolling spread and efficiently grab ballyhoo without being hooked. Though considered by some anglers to be a worthy adversary (North Carolina Division of Marine Fisheries awards a citation for catching and releasing a barracuda), these audacious predators have also been known to slice in half a hooked dolphin (mahi mahi) that was being reeled back to the boat. Thus a barracuda can become the target of a slew of epithets from a disgruntled angler.

Cutlassfish (*Trichiurus lepturus*): The most striking features of this fish are (1) its ribbonlike body (it is also called ribbonfish) tapering to a pointed, whiplike tail and (2) its long, barbed fangs in the front of the mouth, four in the upper jaw and two in the lower jaw. This impressive dental work makes short work of any live prey, but it also permits the cutlassfish to latch on to any lures. I have caught them from both pier and surf while targeting Spanish mackerel and bluefish.

For some anglers, the enjoyment of fishing the Outer Banks waters lies in simply feeling the tug of something wild—whether a "keeper" or a "throwback"—and experiencing the excitement of the fight.

Appendix 1 Outer Banks Fishing Piers

Avalon Fishing Pier
Kill Devil Hills, Milepost 6
(252) 441-7494
www.avalonpier.com

Nags Head Fishing Pier
Nags Head, Milepost 11
(252) 441-5141
www.nagsheadpier.com

Outer Banks Fishing Pier
S. Nags Head, Milepost 18
(252) 441-5740
http://fishingunlimited.net/
    OuterBanksPier.html

Hatteras Island Fishing Pier
Rodanthe, approximately 20 miles
    south of entrance to Cape Hatteras
    National Seashore
(252) 987-2323
www.hatterasislandresort.com

Avon Fishing Pier
Avon, approximately 32 miles
    south of entrance to Cape Hatteras
    National Seashore
(252) 995-5480
www.avonpier.com

Appendix 2 Bait & Tackle Shops

TW's Bait and Tackle
Corolla, 815 Ocean Trail
(252) 433-3339
www.twstackle.com/store_corolla.php

Corolla Bait and Tackle
Corolla, Light Tower Center
(252) 453-9500
www.corollabaitandtackle.com

TW's Bait and Tackle
Kitty Hawk, Milepost 4
(252) 261-7848
www.twstackle.com/store_kitty_
    hawk.php

TI's Bait and Tackle
Nags Head, Milepost 9
(252) 441-4807

TW's Bait and Tackle
Nags Head, Milepost 10
(252) 441-4807
www.twstackle.com/store_nags_
    head.php

Fishing Unlimited
Nags Head Causeway
(252) 441-7413
http://fishingunlimited.net/

Whalebone Tackle
Nags Head Causeway
(252) 441-7413
www.whalebonetackle.com

Hatteras Jack Bait and Tackle
Rodanthe, approximately 19 miles
    south of entrance to Cape Hatteras
    National Seashore
(252) 987-2428
www.hatterasjack.com

Fishing Hole
Salvo, 22 miles south of entrance to
    Cape Hatteras National Seashore

Frank and Fran's: The Fisherman's Friend
Avon, 35 miles south of entrance to
    Cape Hatteras National Seashore
(252) 995-4171
www.hatteras-island.com

Red Drum Tackle
Buxton, in Red Drum shopping center
(252) 995-5414
www.reddrumtackle.com

Dillons Corner
Buxton, across from the Red Drum
(252) 995-5083
www.dillonscorner.com

Frisco Rod and Gun
across from Ramp 49
(252) 995-5366
www.friscorodgun.com

Tradewinds Bait and Tackle
Ocracoke Island, approximately 66 miles
   south of entrance to Cape Hatteras
   National Seashore
(252) 928-5491
www.fishtradewinds.com

Capt Joe Shute's Bait and Tackle
601-H Atlantic Beach Causeway,
   Atlantic Beach
(800) 868-0941
www.captjoes.com

Appendix 3 Marinas

Pirates Cove Marina: sound/inshore/
offshore charters
on the west side of Roanoke Sound
(252) 473-3906 or (800) 367-4728
www.fishpiratescove.com

Broad Creek Fishing Center: sound/
inshore/wrecks/offshore charters
at Wanchese Seaford Industrial Park
(252) 473-9991
www.broadcreekfishingcenter.com

Oregon Inlet Fishing Center: sound/
inshore/offshore charters
4 miles south of entrance to Cape
Hatteras National Seashore
(252) 441-6301 or (800) 272-5199
www.oregon-inlet.com

Oden's Dock: inshore/offshore charters
in Hatteras village, 51 miles south
of entrance to Cape Hatteras
National Seashore
(252) 986-2555
www.odensdock.com

Hatteras Harbor Marina: sound/
inshore/offshore charters
in Hatteras village, 52 miles south
of entrance to Cape Hatteras
National Seashore
(252)986-2166 or (800)676-4939
www.hatterasharbor.com

Teach's Lair Marina: sound/inshore/
offshore charters
south end of Hatteras village, 53 miles
south of entrance to Cape Hatteras
National Seashore
(252) 986-2460 or (888) 868-2460
www.teachslair.com

Hatteras Landing Marina: sound/
inshore/wreck/offshore charters
south end of Hatteras Island, 55 miles
south of entrance to Cape Hatteras
National Seashore
(252) 986-2077 or (800) 551-8478
www.hatteraslanding.com/marina/
fleet.htm

Anchorage Inn and Marina: sound/wreck/
offshore charters
south end of Ocracoke Island, 70 miles
south of entrance to Cape Hatteras
National Seashore
(252) 928-6661
www.theanchorageinn.com/marina.php

Harkers Island Fishing Center:
inshore charters
(252) 728-3907
www.harkersmarina.com

Appendix 4 Kayak Outfitters

Outer Banks Fishing School (Kitty Hawk)
(252) 255-2004
www.obxfishingschool.com

Kitty Hawk Kites
(877) 359-8447
www.kittyhawk.com/kayaking

Index

About the Author

A native of New York City, Stan Ulanski learned from his father how to fish for carp on the reservoirs of upstate New York. From these early experiences arose a lifelong interest in fishing, which now ranges from casting a fly to wary brown trout on spring creeks to tangling with hard-charging blue marlin on the Gulf Stream. Stan's keen interest in the aquatic environment and its inhabitants spills over into his classes at James Madison University, where he teaches courses in oceanography and management of marine resources. He specifically developed a course titled Fly-Fishing Science, which melds the recreational activity of angling with its scientific underpinnings. He is author of *The Gulf Stream: Tiny Plankton, Giant Bluefin, and the Amazing Story of the Powerful River in the Atlantic* (also published by UNC Press) and *The Science of Fly-Fishing* and has done numerous interviews about marine-related issues on National Public Radio programs, including *Living on Earth*.

Though Stan has fished all along the Atlantic and Gulf coasts, he calls the Outer Banks his "home waters," where, in addition to fishing, he enjoys kayaking, swimming, and boating. When not on the Outer Banks, he can be found hiking or bicycling the scenic Shenandoah Valley or using his home in Deltaville, Virginia, as a base to access the Chesapeake Bay.

Other **Southern Gateways Guides** you might enjoy

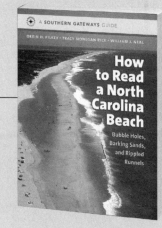

How to Read a North Carolina Beach
Bubble Holes, Barking Sands, and Rippled Runnels

ORRIN H. PILKEY, TRACY MONEGAN RICE, AND WILLIAM J. NEAL

A beachcomber's guide to curiosities along the shore

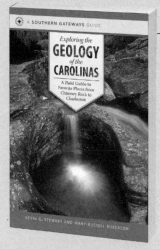

Exploring the Geology of the Carolinas
A Field Guide to Favorite Places from Chimney Rock to Charleston

KEVIN G. STEWART AND MARY-RUSSELL ROBERSON

How to read the rocks of the Carolinas

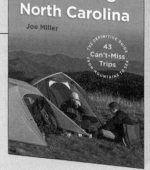

Backpacking North Carolina The Definitive
Guide to 43 Can't-Miss Trips from Mountains to Sea

JOE MILLER

From classic mountain trails to little-known gems of the Piedmont and coastal regions